Meaning in Mathematics

Meaning in Mathematics

Edited by
John Polkinghorne

OXFORD
UNIVERSITY PRESS

OXFORD
UNIVERSITY PRESS

Great Clarendon Street, Oxford OX2 6DP

Oxford University Press is a department of the University of Oxford.
It furthers the University's objective of excellence in research, scholarship,
and education by publishing worldwide in

Oxford New York

Auckland Cape Town Dar es Salaam Hong Kong Karachi
Kuala Lumpur Madrid Melbourne Mexico City Nairobi
New Delhi Shanghai Taipei Toronto

With offices in

Argentina Austria Brazil Chile Czech Republic France Greece
Guatemala Hungary Italy Japan Poland Portugal Singapore
South Korea Switzerland Thailand Turkey Ukraine Vietnam

Oxford is a registered trade mark of Oxford University Press
in the UK and in certain other countries

Published in the United States
by Oxford University Press Inc., New York

© Oxford University Press 2011

The moral rights of the authors have been asserted
Database right Oxford University Press (maker)

First published 2011

British Library Cataloguing in Publication Data

Data available

Library of Congress Cataloging in Publication Data
Library of Congress Control Number: 2011920646

Typeset by SPI Publisher Services, Pondicherry, India
Printed in Great Britain
on acid-free paper by
Clays Ltd, St Ives plc

ISBN 978–0–19–960505–7

1 3 5 7 9 10 8 6 4 2

In grateful memory of Peter Lipton, scholar and friend.

Contents

List of contributors — ix

Introduction — 1
John Polkinghorne

1 **Is mathematics discovered or invented?** — 3
Timothy Gowers

Comment — 13
Gideon Rosen

2 **Exploring the mathematical library of Babel** — 17
Marcus du Sautoy

Comment — 26
Mark Steiner

3 **Mathematical reality** — 27
John Polkinghorne

Comment — 35
Mary Leng

Reply — 39
John Polkinghorne

4 **Mathematics, the mind, and the physical world** — 41
Roger Penrose

Comment — 46
Michael Detlefsen

5 **Mathematical understanding** 49
Peter Lipton

Addendum 55
Stewart Shapiro

6 **Creation and discovery in mathematics** 61
Mary Leng

Comment 70
Michael Detlefsen

7 **Discovery, invention and realism: Gödel and others
on the reality of concepts** 73
Michael Detlefsen

Comment 95
John Polkinghorne

8 **Mathematics and objectivity** 97
Stewart Shapiro

Comment 109
Gideon Rosen

Reply 112
Stewart Shapiro

9 **The reality of mathematical objects** 113
Gideon Rosen

Comment 132
Timothy Gowers

10 **Getting more out of mathematics than what we put in** 135
Mark Steiner

Comment 144
Marcus du Sautoy

References 147

Index 153

List of contributors

Editor: **John Charlton Polkinghorne**, **KBE**, **FRS**, the former president of Queens' College, Cambridge, and the winner of the 2002 Templeton Prize, has been a leading figure in the dialogue of science and religion for more than two decades. He resigned his professorship of mathematical physics at Cambridge University to take up a new vocation in mid-life and was ordained a priest in the Church of England in 1982. A fellow of the Royal Society, he was knighted by Queen Elizabeth II in 1997. In addition to an extensive body of writing on theoretical elementary particle physics, including *Quantum Theory: A Very Short Introduction* (2002), he is the editor or co-editor of four books, the co-author (with Michael Welker) of *Faith in the Living God: A Dialogue* (2001), and the author of nineteen other books on the interrelationship of science and theology, including *Belief in God in an Age of Science* (1998), a volume composed of his Terry Lectures at Yale University, *Science and Theology* (1998), *Faith, Science and Understanding* (2000), *Traffic in Truth: Exchanges between Theology and Science* (2001), *The God of Hope and the End of the World* (2002), *Living with Hope* (2003), *Science and the Trinity: The Christian Encounter with Reality* (2004), *Exploring Reality: The Intertwining of Science and Religion* (2005), *Quantum Physics and Theology: An Unexpected Kinship* (2007), *From Physicist to Priest* (2007), *Theology in the Context of Science* (2008), and *Questions of Truth: Fifty-one Responses to Questions about God, Science and Belief* (2008) with Nicholas Beale.

Michael Detlefsen is McMahon-Hank Professor of Philosophy at the University of Notre Dame and Distinguished Invited Professor of Philosophy at both the University of Paris 7-Diderot and the University of Nancy 2. He has held a senior chaire d'excellence of the ANR in France since 2007. His chief scholarly interests are in the history and philosophy of mathematics and logic. His current projects include a book on Gödel's incompleteness theorems with Timothy McCarthy and various other projects concerning ideals of proof in mathematics.

Marcus du Sautoy is professor of mathematics and Simonyi Professor for the Public Understanding of Science at Oxford University, where he is a fellow of New College. His academic work mainly concerns group theory and number theory, and he is widely known for popularizing mathematics. He was awarded the Berwick Prize of the London Mathematical Society in 2001 and the Faraday Prize by the Royal Society in 2009. He has presented numerous series for BBC TV and radio and is the author of three books, *The Music of the Primes* (2003), *Finding Moonshine* (2007), and *The Num8er My5teries: A Mathematical Odyssey through Everyday Life* (2010), for general audiences.

Timothy Gowers, FRS, is Rouse Ball Professor of Mathematics at Cambridge University and a fellow of Trinity College, Cambridge. He received a Fields Medal in 1998 for his research connecting the fields of functional analysis and combinatorics. Earlier, he was awarded the Junior Whitehead Prize by the London Mathematical Society and the European Mathematical Society Prize. A fellow of the Royal Society, he is the author of *Mathematics: A Very Short Introduction* (2002) and the main editor of *The Princeton Companion to Mathematics* (2008). Launched in 2009, his Polymath Project uses the comment functionality of his blog to produce mathematics collaboratively.

Mary Leng is a lecturer in philosophy at the University of Liverpool. Her research focus is the philosophy of mathematics, with particular reference to issues raised by the applicability of mathematics in empirical science. Dr. Leng has been a visiting fellow in the Department of Logic and Philosophy of Science at the University of California at Irvine, and after a postdoctoral fellowship in the humanities at the University of Toronto, she held a research fellowship at St. John's College, Cambridge, for four years, as well as a visiting junior fellowship at the Peter Wall Institute for Advanced Studies at the University of British Columbia. She is the co-editor (with Alexander Paseau and Michael Potter) of *Mathematical Knowledge* (2007), and the author of *Mathematics and Reality* (2010), both of which were published by Oxford University Press.

Peter Lipton was the Hans Rausing Professor and chair of the History and Philosophy of Science at Cambridge University until his death in 2008. He also had been a fellow of King's College, Cambridge. Much of his work concerned explication and inference in science, but his interests extended broadly across many of the major areas of philosophy. A fellow of the Academy of Medical Sciences, he had been consulting editor of *Studies in the History and Philosophy of Science*, the editor of *Theory, Evidence and Explanation* (1995), and editor or co-editor of three special issues of *Studies in the History and Philosophy of Science*. He was the author of *Inference to the Best Explanation* (1991 and 2004).

Roger Penrose, OM, FRS, the Rouse Ball Professor of Mathematics Emeritus at Oxford University and an emeritus fellow of Wadham College, Oxford, is widely acclaimed for his original and broad-based work in mathematical physics, particularly his contributions to general relativity theory, the foundations of quantum theory, and cosmology. He also has written on the link between fundamental physics and human consciousness. A fellow of the Royal

Society, a foreign associate of the National Academy of Sciences, and a fellow of the European Academy of Sciences, Professor Penrose was knighted for his services to science by Queen Elizabeth II in 1994 and awarded Britain's Order of Merit in 2000. He is the author or co-author of ten books, including *The Emperor's New Mind: On Computers, Minds, and the Laws of Physics* (1989), winner of the 1990 Science Book Prize, *Shadows of the Mind: A Search for the Missing Science of Consciousness* (1994), and *The Large, the Small and the Human Mind* (1997). In addition to his books on consciousness, others he has written for more general audiences include one with Stephen Hawking entitled *The Nature of Space and Time* (1996), *The Road to Reality: A Complete Guide to the Laws of the Universe* (2004), and *Cycles of Time: An Extraordinary New View of the Universe* (2010).

Gideon A. Rosen, Stuart Professor of Philosophy and chair of the Council of the Humanities at Princeton University, specializes in metaphysics, epistemology, philosophy of mathematics, and moral philosophy. He has been a visiting professor at the University of Auckland in New Zealand and held a Mellon Foundation New Directions Fellowship at New York University Law School where he served as the Hauser Fellow in Global Law. Professor Rosen is the author (with John P. Burgess) of *A Subject with No Object: Strategies for Nominalist Reconstrual in Mathematics* (1997), published by Oxford University Press.

Stewart D. Shapiro is O'Donnell Professor of Philosophy at The Ohio State University and professorial fellow at the University of St. Andrews. His research and writing have focused primarily on the philosophy of mathematics, logic, the philosophy of logic, and the philosophy of language. The recipient of several fellowships awarded by the National Endowment for the Humanities and the American Council of Learned Societies, he also has received an Ohio State Award for Scholarly Achievement and an Ohio State University Distinguished Scholar Award. Professor Shapiro was an editor of the *Journal of Symbolic Logic* and has edited five special issues of journals and three books, including the *Oxford Handbook of the Philosophy of Logic and Mathematics* (2005). He is also the author of four Oxford University Press books: *Foundations without Foundationalism: A Case for Second-Order Logic* (1991 and 2000), *Philosophy of Mathematics: Structure and Ontology* (1997 and 2000), *Thinking about Mathematics: The Philosophy of Mathematics* (2000), and *Vagueness in Context* (2006). He is writing a new textbook for Oxford University Press tentatively entitled *Logic for Philosophers*.

Mark Steiner is a professor of philosophy at The Hebrew University of Jerusalem. He has specialized in the philosophy of mathematics as part of his more general attention to the philosophy of science. His work has included a critical account of Ludwig Wittgenstein's philosophy of mathematics. He is the author of *Mathematical Knowledge* (1975) and *The Application of Mathematics as a Philosophical Problem* (1998). His translation from Yiddish into English of *Emune un Apikorses* (1948) by Reuven Agushewitz, a Lithuanian-born Talmudic scholar who attacked the philosophy of materialism in all its historical versions, was published as *Faith and Heresy* (2006). He is now working on a translation of Hume's *Treatise of Human Nature* into Hebrew.

Introduction

John Polkinghorne

Is mathematics a highly sophisticated intellectual game in which the adepts display their skill by tackling invented problems, or are mathematicians engaged in acts of discovery as they explore an independent realm of mathematical reality? Why does this seemingly abstract discipline provide the key to unlocking the deep secrets of the physical universe? How one answers these questions will significantly influence metaphysical thinking about reality. An interdisciplinary Symposium composed of mathematicians, physicists and philosophers met twice, at Castel Gandolfo and in Cambridge, to address these issues. This volume presents the considered form of the contributions that each participant made to the vigorous discussions that took place. Every effort has been made to strike a balance between the precision of thought required for such an enterprise and a reasonable degree of accessibility for a non-specialist reader prepared to make some effort to engage with the issues.

Peter Lipton, Professor of the Philosophy of Science at Cambridge University, was a valued contributor to our first meeting, and we were all greatly saddened by his untimely death before we met again. It is the unanimous wish of all of us involved in the project to dedicate this book to the grateful memory of a fine scholar and a courteous and stimulating colleague.

The first two chapters are written by mathematicians, Timothy Gowers and Marcus du Sautoy. They are able to draw on their long and fruitful experience of doing mathematics. Gowers pays particular attention to how the words 'invention' and 'discovery' are actually used in the mathematical community. He concludes that 'discovery' seems appropriate when there is essentially a sole route of argument leading to a significant conclusion, while 'invention' is preferred when several distinct lines of argument are available. du Sautoy describes an incident of insight arising in a flash of inspiration, an experience which he finds carries the conviction that what had been discerned was 'already there', waiting to be found.

The next two chapters are written by mathematical physicists, John Polkinghorne and Roger Penrose. Polkinghorne seeks to defend mathematical realism by a variety of arguments, ranging from Gödelian incompleteness to the evolution of hominid mathematical ability. Both physicists attribute significance to the role that mathematics plays in affording a strategy for discovery in their subject. Penrose appeals to Gödelian incompleteness as signifying that conscious thought is more than neural computation.

The remaining chapters are written by philosophers. Peter Lipton's chapter is sadly confined to the short paper that he contributed to the initial meeting of the Symposium. It discusses the concepts of knowledge, understanding and explanation, and emphasises the differences he sees between their application in science and in mathematics. Stewart Shapiro helpfully provides an Addendum indicating some ways in which this discussion might be further amplified. Mary Leng denies that the feeling of discovery to which so many mathematicians testify necessarily leads to a Platonic view of mathematical reality. Instead, she suggests it can be understood as arising from the exploration of logical necessity. Michael Detlefson gives an extensive survey of approaches, both ancient and modern, to the debate about invention or discovery. He offers a careful critique of Kurt Gödel's famous analogy of mathematical 'perception' to sense perception. Stewart Shapiro considers the argument that mathematics is a human activity, deriving its conventions from human choices. A key concept for him is 'cognitive command', illustrated by the phenomenon of the necessary agreement between the results of different persons doing the same calculation. He sees this as an encouragement to taking the discovery point of view. Gideon Rosen explores the idea that the status of mathematics corresponds to what he calls 'qualified realism'. He characterizes this judgement as a verdict amounting to mathematics being 'metaphysically second rate', because of its dependence on more fundamental logical facts. Finally, Mark Steiner points us to Descartes rather than Plato. He stresses the fact that mathematics seems to offer 'surplus value', allowing mathematicians to get out of the axioms more than seems to have been put in. (Mathematicians themselves call this the quality of being 'deep'.)

A feature of the Symposium was the liveliness and penetration of its discussions. The participants wish to convey to the readers of this book something of the flavour of this experience, and so each has formulated a brief comment to be attached to the chapter of one of the other participants. We believe that these responses are an important part of reporting what was a stimulating and challenging Symposium.

The meetings of the Symposium were supported by the John Templeton Foundation and all participants wish to express their gratitude for this generosity. In particular, we were greatly helped by the organizing skill and keen interest of Dr. Mary Ann Meyers of the Foundation, to whom we offer our special thanks.

1
Is mathematics discovered or invented?

Timothy Gowers

The title of this chapter is a famous question. Indeed, perhaps it is a little too famous: it has been asked over and over again, and it is not clear what would constitute a satisfactory answer. However, I was asked to address it during the discussions that led to this volume, and since most of the participants in those discussions were not research mathematicians, I was in particular asked to give a mathematician's perspective on it.

One reason for the appeal of the question seems to be that people can use it to support their philosophical views. If mathematics is discovered, then it would appear that there is something out there that mathematicians are discovering, which in turn would appear to lend support to a Platonist conception of mathematics, whereas if it is invented, then that might seem to be an argument in favour of a non-realist view of mathematical objects and mathematical truth.

But before a conclusion like that can be drawn, the argument needs to be fleshed out in detail. First, one must be very clear what it means to say that some piece of mathematics has been discovered, and then one must explain, using that meaning, why a Platonist conclusion follows. I do not myself believe that this programme can be carried out, but one can at least make a start on it by trying to explain the incontestable fact that almost all mathematicians who successfully prove theorems feel as though they are making discoveries. It is possible to think about this question in a non-philosophical way, which is what I shall try to do. For instance, I shall consider whether there is an identifiable distinction between parts of mathematics that feel like discoveries and parts that feel like inventions. This is partly a psychological question and partly a question about whether there are objective properties of mathematical statements that explain how they are perceived. The argument in favour of Platonism only needs *some* of mathematics to be discovered: if it turns out that there are two broad kinds of mathematics, then perhaps one can understand the distinction and formulate more precisely what mathematical discovery (as opposed to the mere producing of mathematics) is.

As the etymology of the word 'discover' suggests, we normally talk of discovery when we find something that was, unbeknownst to us, already there. For example, Columbus is said to have discovered America (even if one can question that statement for other reasons), and Tutankhamun's tomb was discovered by Howard Carter in 1922. We say this even when we cannot directly observe what has been discovered: for instance, J. J. Thompson is famous as the discoverer of the electron. Of greater relevance to mathematics is the discovery of facts: we discover *that* something is the case. For example, it would make perfectly good sense to say that Bernstein and Woodward discovered (or contributed to the discovery) that Nixon was linked to the Watergate burglary.

In all these cases, we have some phenomenon, or fact, that is brought to our attention by the discovery. So one might ask whether this transition from unknown to known could serve as a definition of discovery. But a few examples show that there is a little more to it than that. For instance, an amusing fact, known to people who like doing cryptic crosswords, is that the words 'carthorse' and 'orchestra' are anagrams. I presume that somebody somewhere was the first person to notice this fact, but I am inclined to call it an observation (hence my use of the word 'notice') rather than a discovery. Why is this? Perhaps it is because the words 'carthorse' and 'orchestra' were there under our noses all the time and what has been spotted is a simple relationship between them. But why could we not say that the relationship is discovered even if the words were familiar? Another possible explanation is that once the relationship is pointed out, one can very easily verify that it holds: you don't have to travel to America or Egypt, or do a delicate scientific experiment, or get access to secret documents.

As far as evidence for Platonism is concerned, the distinction between discovery and observation is not especially important: if you notice something, then that something must have been there for you to notice, just as if you discover it then it must have been there for you to discover. So let us think of observation as a mild kind of discovery rather than as a fundamentally different phenomenon.

How about invention? What kinds of things do we invent? Machines are an obvious example: we talk of the invention of the steam engine, or the aeroplane, or the mobile phone. We also invent games: for instance, the British invented cricket—and more to the point, that is an appropriate way of saying what happened. Art supplies us with a more interesting example. One would never talk of a single work of art being invented, but it does seem to be possible to invent a style or a technique. For example, Picasso did not invent *Les Desmoiselles d'Avignon*, but he and Braque are credited with inventing cubism.

A common theme that emerges from these examples is that what we invent tends not to be individual objects: rather, we invent general *methods* for producing objects. When we talk of the invention of the steam engine, we are not talking about one particular instance of steam-enginehood, but rather of the idea—that a clever arrangement of steam, pistons, etc. can be used to drive machines—that led to the building of many steam engines. Similarly, cricket is a set of rules that has led to many games of cricket, and cubism is a general idea that led to the painting of many cubist pictures.

If somebody wants to argue that the fact of mathematical discovery is evidence for a Platonist view of mathematics, then what they will be trying to show is that certain abstract entities have an independent existence, and certain facts about those entities are true for much the same sort of reason that certain facts about concrete entities are true. For instance, the statement 'There are infinitely many prime numbers' is true, according to this view, because there really are infinitely many natural numbers out there, and it really is the case that infinitely many of them are prime.

A small remark one could make here is that it is also possible to use the concept of invention as an argument in favour of an independent existence for abstract concepts. Indeed, our examples of invention all involve abstraction in a crucial way: 'the steam engine', as we have just noted, is an abstract concept, as are the rules of cricket. Cubism is a more problematic example as it is less precisely defined, but it is undoubtedly abstract rather than concrete. Why do we not say that these abstract concepts are brought into existence when we invent them?

One reason is that we feel that independently existing abstract concepts should be timeless. So we do not like the idea that when the British invented the rules of cricket, they reached out into the abstract realm and brought the rules into existence. A more appealing picture would be that they selected the rules of cricket from a vast 'rule space' that consists of all possible sets of rules (most of which give rise to terrible games). A drawback with this second picture is that it fills up the abstract realm with a great deal of junk, but perhaps it really is like that. For example, it is supposed to contain all the real numbers, all but countably many of which are undefinable.

Another argument against the idea that one brings an abstract concept into existence when one invents it is that the concepts that we invent are not fundamental enough: they tend to be methods for dealing with other objects, either abstract or concrete, that are much simpler. For example, the rules of cricket describe constraints on a set of procedures that are carried out by 22 players, a ball and two wickets. From an ontological point of view, the players, ball and wickets seem more secure than the constraints on how they behave.

Earlier, I commented that we do not normally talk of inventing a single work of art. However, we do not discover it either: a commonly used word for what we do would be 'create'. And most people, if asked, would say that this kind of creation has more in common with invention than with discovery, just as observation has more in common with discovery than with invention.

Why is this? Well, in both cases what is brought into existence has many arbitrary features: if we could turn the clock back to just before cricket was invented and run the world all over again, it is likely that we would see the invention of a similar game, but unlikely that its rules would be identical to those of the actual game of cricket. (One might object that if the laws of physics are deterministic, then the world would develop precisely as it did the first time. In that case, one could make a few small random changes before the rerun.) Similarly, if somebody had accidentally destroyed *Les Desmoiselles d'Avignon* just after Picasso started work on it, forcing him to start again, it is likely that he would have produced a similar but perceptibly different painting. By contrast, if Columbus had not existed, then somebody else would have discovered *America* and not just some huge land mass of a broadly similar kind on the other side of the Atlantic. And the fact that 'carthorse' and 'orchestra' are anagrams is independent of who was the first to observe it.

With these thoughts in mind, let us turn to mathematics. Again, it will help to look at some examples of what people typically say about various famous parts of the subject. Let me list some discoveries, some observations and some inventions. (I cannot think of circumstances where I would definitely want to say that a piece of mathematics was created.) Later I will try to justify why each item is described in the way it is.

A few well-known discoveries are the formula for the quadratic, the absence of a similar formula for the quintic, the Monster group, and the fact that there are infinitely many primes. A few observations are that the number of primes less than 100 is 25, that the last digits of the powers of 3 form the sequence $3, 9, 7, 1, 3, 9, 7, 1, \ldots$, and that the number 10001 factorizes as 73 times 137. An intermediate case is the fact that if you define an infinite sequence $z_0, z_1, z_2, \ldots$ of complex numbers by setting $z_0 = 0$ and $z_n = z_{n-1}^2 + C$ for every $n > 0$, then the set of all complex numbers C for which the sequence does not tend to infinity, now called the Mandelbrot set, has a remarkably complicated structure. (I regard this as intermediate because, although Mandelbrot and others stumbled on it almost by accident, it has turned out to be an object of fundamental importance in the theory of dynamical systems.)

On the other side, it is often said that Newton and Leibniz independently invented the calculus. (I planned to include this example, and was heartened when, quite by coincidence, on the day that I am writing this paragraph, there was a plug for a radio programme about their priority dispute, and the word 'invented' was indeed used.) One also sometimes talks of mathematical theories (as opposed to theorems) being invented: it does not sound ridiculous to say that Grothendieck invented the theory of schemes, though one might equally well say 'introduced' or 'developed'. Similarly, any of these three words would be appropriate for describing what Cohen did to the method of forcing, which he used to prove the independence of the continuum hypothesis. From our point of view, what is interesting is that the words 'invent', 'introduce' and 'develop' all carry with them the suggestion that some general technique is brought into being.

A mathematical object about which there might be some dispute is the number i, or more generally the complex number system. Were complex numbers discovered or invented? Or rather, would mathematicians normally refer to the arrival of complex numbers into mathematics using a discovery-type word or an invention-type word? If you type the phrases 'complex numbers were invented' and 'complex numbers were discovered' into Google, you get approximately the same number of hits (between 4500 and 5000 in both cases), so there appears to be no clear answer. But this too is a useful piece of data. A similar example is non-Euclidean geometry, though here 'discovery of non-Euclidean geometry' outnumbers 'invention of non-Euclidean geometry' by a ratio of about 3 to 1.

Another case that is not clear-cut is that of *proofs*: are they discovered or invented? Sometimes a proof seems so natural—mathematicians often talk of 'the right proof' of a statement, meaning not that it is the only correct proof but that it is the one proof that truly explains why the statement is true—that the word 'discover' is the obvious word to use. But sometimes it feels more appropriate to say something like, 'Conjecture 2.5 was first proved in 1990, but in 2002 Smith came up with an ingenious and surprisingly short argument that actually establishes a slightly more general result.' One could say 'discovered' instead of 'came up with' in that sentence, but the latter captures better the idea that Smith's argument was just one of many that there might have been, and that Smith did not simply stumble on it by accident.

Let us take stock at this point, and see whether we can explain what it is about a piece of mathematics that causes us to put it into one of the three categories: discovered, invented, or not clearly either.

The non-mathematical examples suggested that discoveries and observations were usually of objects or facts over which the discoverer had no control, whereas inventions and creations were of objects or procedures with many features that could be chosen by the inventor or creator. We also drew some more refined, but less important, distinctions within each class. A discovery tended to be more notable than an observation and less easy to verify afterwards. And inventions tended to be more general than creations.

Do these distinctions continue to hold in much the same form when we come to talk about mathematics? I claimed earlier that the formula for the quadratic was discovered, and when I try out the phrase 'the invention of the formula for the quadratic', I find that I do not like it, for exactly the reason that the solutions of $ax^2 + bx + c$ are the numbers $(-b \pm \sqrt{b^2 - 4ac})/2a$. Whoever first derived that formula did not have any choice about what the formula would eventually be. It is of course possible to *notate* the formula differently, but that is another matter. I do not want to get bogged down in a discussion of what it means for two formulae to be 'essentially the same', so let me simply say that the formula itself was a discovery but that different people have *come up with* different ways of expressing it. However, this kind of concern will reappear when we look at other examples.

The insolubility of the quintic is another straightforward example. *It is* insoluble by radicals, and nothing Abel did could have changed that. So his famous theorem was a discovery. However, aspects of his *proof* would be regarded as invention—there have subsequently been very different looking proofs. This is particularly clear with the closely related work of Galois, who is credited with the invention of group theory. (The phrase 'invention of group theory' has 40,300 entries in Google, compared with 10 for 'discovery of group theory'.)

The Monster group is a more interesting case. It first entered the mathematical scene when Fischer and Griess predicted its existence in 1973. But what does that mean? If they could refer to the Monster group at all, then does that not imply that it existed? The answer is simple: they predicted that a group with certain remarkable *properties* (one of which is its huge size—hence the name) existed and was unique. So to say 'I believe that the Monster group exists' was shorthand for 'I believe that there exists a group with these amazing properties' and the name 'Monster group' was referring to a hypothetical entity.

The existence and uniqueness of the Monster group were indeed proved, though not until 1982 and 1990, respectively, and it is not quite clear whether we should regard this mathematical advance as a discovery or an invention. If we ignore the story and condense 17 years to an instant, then it is tempting to say that the Monster group was there all along until it was discovered by group theorists. Perhaps one could even add a little detail: back in 1973 people started to have reason to suppose that it existed, and they finally bumped into it in 1982.

But how did this 'bumping' take place? Griess did not prove in some indirect way that the Monster group had to exist (though such proofs are possible in mathematics). Rather, he *constructed* the group. Here, I am using the word that all mathematicians would use. To construct it, he constructed an auxiliary object, a complicated algebraic structure now known as the Griess algebra, and showed that the symmetries of this algebra formed a group with the desired properties. However, this is not the only way of obtaining the Monster group: there are other constructions that give rise to groups that have the same properties, and hence, by the uniqueness result, are isomorphic to it. So it seems that Griess had some control over the process by which he built the Monster group, even if what he ended up building was determined in advance. Interestingly, the phrase 'construction of the Monster group' is much more popular on Google than the phrase 'discovery of the Monster group' (8290 to 9), but if you change it to 'the construction of the Monster group' then it becomes much less popular (6 entries), reflecting the fact that there are many different constructions.

Another question one might ask is this. If we do decide to talk about the discovery of the Monster group, are we talking about the discovery of an *object*, the Monster group, or of a *fact*, the fact that there exists a group with certain properties and that that group is unique? Certainly, the second is a better description of the work that the group theorists involved actually did, and the word 'construct' is a better word than 'discover' at describing how they proved the existence part of this statement.

The other discoveries and observations listed earlier appear to be more straightforward, so let us turn to the examples on the invention side.

A straightforward use of the word 'invention' in mathematics is to refer to the way general theories and techniques come into being. This certainly covers the example of the calculus, which is not an object, or a single fact, but rather a large collection of facts and methods that greatly increase your mathematical power when you are familiar with them. It also covers Cohen's technique of forcing: again, there are theorems involved, but what is truly interesting about forcing is that it is a general and adaptable method for proving independence statements in set theory.

I suggested earlier that inventors should have some control over what they invent. That applies to these examples: there is no clear criterion that says which mathematical statements are part of the calculus, and there are many ways of presenting the theory of forcing (and, as I mentioned earlier, many generalizations, modifications and extensions of Cohen's original ideas).

How about the complex number system? At first sight this does not look at all like an invention. After all, it is provably unique (up to the isomorphism that sends $a + bi$ to $a - bi$), and it is an object rather than a theory or a technique. So why do people sometimes call it an invention, or at the very least feel a little uneasy about calling it a discovery?

I do not have a complete answer to this question, but I suspect that the reason it is a somewhat difficult example is similar to the reason that the Monster group is difficult, which is that one can 'construct' the complex numbers in more than one way. One approach is to use something like the way they were constructed historically (my knowledge of the history is very patchy, so I shall not say *how* close the resemblance is). One simply introduces a new symbol, i, and declares that it behaves much like a real number, obeying all the usual algebraic rules, and has the additional property that $i^2 = -1$. From this one can deduce that

$$(a + bi)(c + di) = ac + bci + adi + bdi^2 = (ad - bd) + (ad + bc)i$$

and many other facts that can be used to build up the theory of complex numbers. A second approach, which was introduced much later in order to demonstrate that the complex number system was consistent if the real number system was, is to define a complex number to be an ordered pair (a, b) of real numbers, and to stipulate that addition and multiplication of these ordered pairs are given by the following rules:

$$(a, b) + (c, d) = (a + c, b + d)$$
$$(a, b)(c, d) = (ac - bd, ad + bc)$$

This second method is often used in university courses that build up the number systems rigorously. One proves that these ordered pairs form a field under the two given operations, and finally one says, 'From now on I shall write $a + bi$ instead of (a, b).'

Another reason for our ambivalence about the complex numbers is that they feel less real than real numbers. (Of course, the names given to these numbers reflect this rather unsubtly.) We can directly relate the real numbers to quantities such as time, mass, length, temperature, and so on (though for this we never need the infinite precision of the real number system), so it feels as though they have an independent existence that we observe. But we do not run into the complex numbers in that way. Rather, we play what feels like a sort of game—imagine what would happen if -1 *did* have a square root.

But why in that case do we not feel happy just to say that the complex numbers were invented? The reason is that the game is much more interesting than we had any right to expect, and it has had a huge influence even on those parts of mathematics that are about real numbers or even integers. It is as though after our one small act of inventing i, the game took over and we lost control of the consequences. (Another example of this phenomenon is Conway's famous game of Life. He devised a few simple rules, by a process that one would surely want to regard as closer to invention than discovery, but once he had done so he found that he had created a world full of unexpected phenomena that he had not put there, so to speak. Indeed, most of them were discovered—to use the obvious word—by other people.)

Why is 'discovery of non-Euclidean geometry' more popular than 'invention of non-Euclidean geometry'? This is an interesting case, because there are two approaches to the subject, one axiomatic and one concrete. One could talk about non-Euclidean geometry as the discovery of the remarkable fact that a different set of axioms, where the parallel postulate is replaced by a statement that allows a line to have several parallels through any given point, is consistent. Alternatively, one could think of it as the construction of models in which those axioms are true. Strictly speaking, one needs the second for the first, but if one explores in detail the consequences of the axioms and proves all sorts of interesting theorems without ever reaching a contradiction, that can be quite impressive evidence for their consistency. It is probably because the consistency interests us more than the particular choice of model, combined with the fact that any two models of the hyperbolic plane are isometric, that we usually call it a discovery. However, Euclidean geometry (wrongly) feels more 'real' than hyperbolic geometry, and there is no single model of hyperbolic geometry that stands out as the most natural one; these two facts may explain why the word 'invention' is sometimes used.

My final example was that of proofs, which I claimed could be discovered *or* invented, depending on the nature of the proof. Of course, these are by no means the only two words or phrases that one might use: some others are 'thought of', 'found', 'came up with'. And often one regards the proof less as an object than as a process, and focuses on what is proved, as is shown in sentences such as, 'After a long struggle, they eventually managed to prove/establish/show/demonstrate that ...' Proofs illustrate once again the general point that we use discovery words when the author has less control and invention words when there are many choices to be made. Where, one might ask, does the choice come from? This is a fascinating question in itself, but let me point out just one source of choice and arbitrariness: often a proof requires one to show that a certain mathematical object or structure exists (either as the main statement or as some intermediate lemma), and often the object or structure in question is far from unique.

Before drawing any conclusions from these examples, I would like to discuss briefly another aspect of the question. I have been looking at it mainly from a linguistic point of view, but, as I mentioned right at the beginning, it also has a strong psychological component: when one is doing mathematical research, it sometimes feels more like discovery and sometimes more like invention. What is the difference between the two experiences?

Since I am more familiar with myself than with anybody else, let me draw on my own experience. In the mid 1990s I started on a research project that has occupied me in one way or another ever since. I was thinking about a theorem that I felt ought to have a simpler proof than the two that were then known. Eventually I found one (here I am using the word that comes naturally); unfortunately it was not simpler, but it gave important new information. The process of finding this proof felt much more like discovery than invention, because by the time I reached the end, the structure of the argument included many elements that I had not even begun to envisage when I started working on it. Moreover, it became clear that there was a large body of closely related facts that added up to a coherent and yet-to-be-discovered theory. (At this stage, they were not proved facts, and not always even precisely stated facts. It was just clear that 'something was going on' that needed to be investigated.) I and several others have been working to develop this theory, and theorems have been proved that would not even have been stated as conjectures fifteen years ago.

Why did this feel like discovery rather than invention? Once again it is connected with control: I was not selecting the facts I happened to like from a vast range of possibilities. Rather, certain statements stood out as obviously natural and important. Now that the theory is more developed, it is less clear which facts are central and which more peripheral, and for that reason the enterprise feels as though it has an invention component as well.

A few years earlier, I had a different experience: I found a counterexample to an old conjecture in the theory of Banach spaces. To do this, I constructed a complicated Banach space. This felt partly like an invention—I did have arbitrary choices, and many other counterexamples have subsequently been found—and partly like a discovery—much of what I did was in response to the requirements of the problem, and felt like the natural thing to do, and a very similar example was discovered independently by someone else (and even the later examples use similar techniques). So this is another complicated situation to analyse, but the reason it is complicated is simply that the question of how much control I had is a complicated one.

What conclusion should we draw from all these examples and from how we naturally seem to regard them? First, it is clear that the question with which we began is rather artificial. For a start, the idea that either all of mathematics is discovered or all of mathematics is invented is ridiculous. But even if we look at the origins of individual pieces of mathematics, we are not forced to use the word 'discover' or 'invent', and we very often don't.

Nevertheless, there does seem to be a spectrum of possibilities, with some parts of mathematics feeling more like discoveries and others more like inventions. It is not always easy to say which are which, but there does seem to be one feature that correlates strongly with whether we prefer to use a discovery-type word or an invention-type word. That feature is the control that we have over what is produced. This, as I have argued, even helps to explain why the doubtful cases are doubtful.

If this is correct (perhaps after some refinement), what philosophical consequences can we draw from it? I suggested at the beginning that the answer to the question did not have any bearing on questions such as 'Do numbers exist?' or 'Are mathematical statements true because the objects they mention really do relate to each other in the ways described?' My reason for that suggestion is that pieces of mathematics have objective features that explain how much control we have over them. For instance, as I mentioned earlier, the proof of an existential statement may well be far from unique, for the simple reason that there may be many objects with the required properties. But this is a straightforward mathematical phenomenon. One could accept my analysis and believe that the objects in question 'really exist', or one could view the statements that they exist as moves in games played with marks on paper, or one could regard the objects as convenient fictions. The fact that some parts of mathematics are unexpected and others not, that some solutions are unique and others multiple, that some proofs are obvious and others take a huge amount of work to produce—all these have a bearing on how we describe the process of mathematical production and all of them are entirely independent of one's philosophical position.

Comment on Timothy Gowers' 'Is mathematics discovered or invented?'

Gideon Rosen

In a nearby possible world Timothy Gowers is not the distinguished mathematician that he is in our world, but rather a 1950s-style ordinary language philosopher. In his contribution to this volume he approaches his title question—'Is mathematics discovered or invented?'—by attending rather carefully to the ways in which mathematicians (and the variously informed hordes whose musings are lodged in Google's database) actually use these words in application to various parts of mathematics. Gowers' conclusion is (roughly) that the rhetoric of 'discovery' strikes us as apt when the mathematician has no significant choice about how he does what he does, whereas we are inclined to speak of 'invention' or perhaps 'construction' when there are many ways to perform the task at hand and the mathematician has some control over how he does it.

Gowers is keen to insist that the distinctions that interest him are independent of one's views about the metaphysics of mathematics.

> One could accept my analysis and believe that the objects in question "really exist", or one could view the statements that they exist as moves in games played with marks on paper, or one could regard the objects as convenient fictions. The fact that some parts of mathematics are unexpected and others not, that some solutions are unique and others multiple, that some proofs are obvious and others take a huge amount of work to produce—all these have a bearing on how we describe the process of mathematical production and all of them are entirely independent of one's philosophical position. (p. 12)

This strikes me as exactly right, but it raises a question that Gowers does not address. Gowers has described the conditions under which mathematicians are inclined to *say* that some achievement amounts to a discovery or an invention, and also the conditions under which an achievement is likely to *feel* like a discovery or an invention to those whose achievement it is. But how seriously should we take these linguistic and psychological observations? As philosophers have often noted, it is one thing to chart the circumstances under which we are inclined to *say* this or that, another to identify the conditions under which it is literally *correct* to say this or that. So let us grant that mathematicians agree in their classification of some episode as a matter of (say) invention. Does that entail or even suggest that this episode was in fact a matter of invention? Or is this rather a mere manner of speaking that it would be a mistake to take too seriously?

I believe that this question has different answers in different cases. As Gowers notes, we speak of the invention/discovery of many kinds of thing: theories, theorems, proofs and proof techniques, but also mathematical objects of various sorts (numbers, number systems). We can say that Cantor invented the theory of transfinite numbers, but we are much less likely to say that Cantor invented the transfinite numbers themselves. Let's focus first on the rhetoric of invention/construction as applied to mathematical objects. Here Gowers discusses the case of the Monster group, an enormous finite group whose existence and uniqueness were proved in 1982 and 1990, respectively. The linguistic evidence suggest that mathematicians are more inclined to speak of the 'construction' of the Monster group than of its 'discovery', and Gowers' account explains this. The proof of the existence of the Monster group is not unique: many examples may be adduced to establish the existential theorem that a group with the relevant properties exists, even though (as it happens) every such example is isomorphic to every other. But is there any reason to take the imagery of construction seriously in this case? In my view it is a non-negotiable feature of the literal use of this idiom that if a thing has been invented or constructed, it did not exist before it was invented and would not have existed if it had not been invented. By contrast, when a thing is discovered, it must exist prior to (or at least independently of) the episode of discovery. But as I think Gowers would agree, it would be quite odd to say that before 1982, the Monster group did not exist. If this were the right thing to say, then when Griess first asked himself the question, 'Does the Monster exist?' the answer should have been obvious: 'Not yet, but maybe someday.' But in fact no one speaks of mathematical objects in this way. I am therefore tempted to conclude that even if Gowers is right about the conditions under which we are inclined to reach for the language of invention or construction in connection with mathematical objects, it would be a mistake to take this language literally in this connection.

Things are otherwise when it comes to mathematical theories—especially large theories like the calculus. If someone had asked in (say) 1650 whether there existed a powerful battery of algebraic techniques for calculating the area bounded by a curve and the line tangent to a curve at a point, and a deep theory justifying these techniques and displaying their connections, the answer might well have been, 'Not yet, but maybe someday.' Moreover, it seems equally natural to say that if no one had ever managed to write down such a theory, then the calculus, as we know it, would not exist. Theories of this sort thus appear to belong to the same ontological category as novels and poems and philosophical treatises. Such things are *abstract artifacts*: abstract entities that come into existence when someone produces a concrete representation of them for the first time. In these cases, I see no reason not to take the rhetoric of invention seriously as a sober and literal account of the underlying metaphysics.

Gowers makes no claims of this sort, but I wonder, however, whether he would agree with my assumption that unless we are prepared to say that the invented item did not exist prior to its invention, we should regard claims of invention (construction, creation, etc.) in mathematics as metaphorical. We might then take Gowers' careful account of the conditions under which we are inclined to deploy the metaphor as an account of the sober and metaphysically neutral truth that underlies it.

2
Exploring the mathematical library of Babel

Marcus du Sautoy

I'm a mathematician, not a philosopher. My job is to prove new theorems. To discover new truths about the numbers we count with. To create new symmetrical objects. To find new connections between disparate parts of the mathematical landscape.

Yet contained in my job description are a whole bunch of words that raise important questions about what mathematics is and how it relates to the physical and mental world we inhabit. 'Create', 'discover', 'proof', 'truth'. All very emotive words. And every mathematician at some point will find themselves contemplating whether a new mathematical breakthrough they've just made is an act of creation or an act of discovery. Is mathematics an objective or subjective activity? Do mathematical objects exist?

The only way for me to engage with these questions is to analyse what I think I do when I do mathematics. So I've chosen an episode from my working life to help me explore some of these issues. (More details of this discovery can be found in du Sautoy, 2009.)

One of my proudest moments as a mathematician was constructing a new symmetrical object whose subgroup structure is related to counting the number of solutions modulo p of an elliptic curve. Finding solutions to elliptic curves is one of the toughest problems on the mathematical books. An elliptic curve is an equation like $y^2 = x^3 - x$ (or more generally a quadratic in y equal to a cubic in x). One of the million-dollar-prize problems offered by the Clay Institute, called the Birch and Swinnerton–Dyer Conjecture, seeks to understand when one of these equations has infinitely many solutions where both x and y are fractions.

I constructed this symmetrical object whose structure encoded this important question of solving equations while working at the Max Planck Institute in Bonn. A mathematical theorem that I'd proved with a colleague in Germany hinted that such symmetrical objects might exist, but until such a symmetry

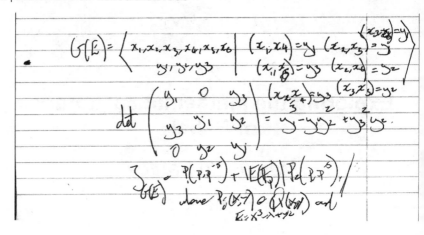

Fig. 2.1 The construction in my notebook of a new symmetrical object.

group was constructed that demonstrated the connection it might just be an illusion. That evening sitting in Bonn was one of those moments that mathematicians often talk about, when a flash of inspiration hits you. I wrote down the construction of how these symmetries of the new object should interact with each other on the yellow legal pad that is the palette for my mathematical musings.

It felt right. It took a few more days to really prove what I thought it did. But once the details were fleshed out, this new object revealed a connection between the world of symmetry and the world of arithmetic geometry that had not been seen before.

Of course, when I say I constructed this symmetrical object, I didn't physically build it. This is an object of the mind living in the abstract world of mathematics. It isn't like the person who first carved out an icosahedron with its 20 triangular faces or the Moorish artist in the Alhambra, Granada, who found a new way to cover the walls with symmetrical tiles. A physical representation of the object I discovered would only exist in some high-dimensional space. And even so, these representations are only expressions of the underlying group of symmetries. Both the rotations of the icosahedron and the dodecahedron are examples of two objects with the same underlying symmetry group called A_5. Similarly, these two designs found in the Alhambra, although physically very different, have identical underlying groups of symmetries.

Just as the number 'three' captures the identity of a collection which has three objects, whether it be three apples or three kangaroos, the naming of the symmetry 632 has abstracted the symmetrical identity shared by these two walls in the Alhambra. The abstract symmetry group is described by giving names to each of the symmetries and then explaining how these symmetries interact with each other when you do one symmetry followed by another.

Fig. 2.2 Two walls in the Alhambra with the same group of symmetries called 632.

What I'd 'constructed' that evening in Bonn was an abstract symmetrical object whose symmetries interact in such a way to produce interesting new connections with elliptic curves. It certainly doesn't exist in the physical world, yet when you spend enough time in the mathematical world it has a reality that is akin to handling a dodecahedron or tiling a wall in the Alhambra.

I've been careful to avoid using the word 'create' while describing the episode above but I had to fight myself not to write the word. Because constructing this new group of symmetries certainly felt like an act of creation. I experienced the strong sense that the scribblings I penned on my yellow pad brought into existence something new, something that didn't exist before I'd described its contours. It was through my act of imagination that this thing emerged. It required my agency to realize the existence of this object. It wasn't something that would naturally evolve without me being present. I provided it with the push that gave it life.

The creative side of mathematics is one that many mathematicians talk about. It is one of the reasons that drew me to mathematics rather than the other sciences, which I felt were more about observation. When I was at school, I was very interested in music; I was learning the trumpet, I enjoyed the theatre, enjoyed reading. Science hadn't really captured my imagination. But then around the age of 13 my mathematics teacher took me aside after one lesson: 'I think you should find out what mathematics is really about. Mathematics is not about all the multiplication tables and long division we do in the classroom. It is actually much more exciting than that and I think you might enjoy seeing the bigger picture.' He gave me the names of some books that he thought I might enjoy and would open up what this world of mathematics was all about.

One of the books was *A Mathematician's Apology* by G. H. Hardy (1940). It had a big impact on me. As I read Hardy's book, there were sentences which revealed to me that mathematics shared a lot in common with the creative arts. It seemed to be compatible with things I loved doing: languages, music, reading. Here for example is Hardy, writing about being a mathematician: 'A mathematician, like a painter or a poet, is a maker of patterns. If his patterns are more permanent than theirs, it is because they are made with *ideas*.' Later he writes: 'The mathematician's patterns, like the painter's or the poet's, must be *beautiful*; the ideas, like the colours or the words, must fit together in an harmonious way. Beauty is the first test: there is no permanent place in the world for ugly mathematics.' For Hardy, mathematics was a creative art, not a useful science. 'The "real" mathematics of the "real" mathematicians, the mathematics of Fermat and Euler and Gauss and Abel and Riemann, is almost wholly "useless" (and this is true of "applied" as of "pure" mathematics). It is not possible to justify the life of any genuine professional mathematician on the ground of the "utility" of his work.'

The construction of my group of symmetries certainly wasn't motivated by utility. It was the creation of something that appealed to my sense of aesthetics. It was surprising. It had an unexpected twist. Like a theme in a piece of music, it mutated during the course of the proof into something quite different. I suppose part of my motivation for constructing this group of symmetries was a certain *mathematical* utility. It might ultimately help us understand elliptic curves better; it gives a new perspective on the complexity of classifying *p*-groups. But it still felt like a creative act not forced on me by external factors outside my control.

And yet... wasn't this mathematical object just sitting out there waiting for someone to notice it? Wasn't my moment in Bonn just an act of discovery? If it wasn't me who discovered it, wouldn't someone else have eventually come to the same realization? I was scrambling round the mathematical landscape and uncovered this symmetrical object. Wasn't it there all along, waiting to be carved out of the ground? Why was it any different to the first scientist who discovered gold or the first astronomer to spot the planet Neptune.

Here's Hardy in a completely different frame of mind from his talk of the creativity of mathematics: 'I believe that mathematical reality lies outside us, that our function is to discover or observe it and that the theorems which we prove and which we describe grandiloquently as our "creations" are simply our notes of our observations.' It sums up the schizophrenic relationship that I think all mathematicians have towards their work. No creative accountancy can make a prime number divisible. As Hardy declared: '317 is a prime not because we think so, or because our minds are shaped in one way or another, but *because it is so*, because mathematical reality is built that way.'

Maybe there is a difference between discovering a new group of symmetries and discovering a new element or planet because gold and Neptune have naturally evolved and didn't require my agency. But I still feel that if I hadn't discovered this symmetry group, then it was lying out there for someone else to construct. How much is it a product of my imagination? History records a catalogue of events where mathematical objects were discovered simultaneously and independently by different mathematicians. The most famous is the discovery of non-Euclidean geometries which was made independently by Gauss, Bolyai and Lobachevsky. Although their notation, descriptions and explanations might have been quite personal, the object they discovered, a geometry with triangles whose angles add up to less than 180 degrees, was the same.

In contrast, one can't imagine three composers simultaneously composing 'The Death and the Maiden' string quartet. That was a creation of Schubert's genius, made at the same time as non-Euclidean geometry was first emerging. But although the piece of music itself is unique and could never be replicated by another composer, it is striking that moods and changes in genre in music and the other arts are happening independently and simultaneously. Composers are discovering new ways of composing, new structures, new possibilities, often at the same time. Schubert's quartet marks the beginning of the Romantic period of musical composition. But he wasn't the only one exploring the ideas of restless key modulations and the heightened contrasts of Sturm und Drang. Contemporary composers I've worked with talk of being beaten to the discovery of an idea, as if composers are equally discovering new structures, new forms within which to frame their composition.

Perhaps I can make a proposal that explains the feeling of creativity that I have when I do mathematics. There were many different groups of symmetries that I could have written down on my yellow pad that evening in Bonn. Infinitely many in fact. All I have to do is write down names for the symmetries and define how they interact and... voilà... I've created/discovered a new group. There will be a question of whether the group of symmetries has been constructed before, but I am more interested in focusing on why I was so excited about the particular group of symmetries I constructed that evening.

I think it is helpful to consider an analogy with a composer or a writer. I can randomly write down notes on a stave, give the notes different lengths, different dynamics, and I will have composed a piece of music. Or I can sit at a typewriter and just bang out strings of letters or words and write a book. Borges's *The Library of Babel* contains every book composed of 25 letters where each book consists of 410 pages; each page is made up of 40 lines each consisting of 80 positions. There are of course a lot of books in the library, $25^{1312000}$ to be precise.

They are all sitting there waiting for an author to discover one. The possibility of *Great Expectations* existed out there before Charles Dickens pulled the book off the shelf. The act of creativity is in singling out this book among all the possible books to write. And I think that it is the same characteristic that is involved in doing mathematics and which is often overlooked.

I could write down endless new and original theorems. I could construct infinitely many new symmetry groups. I could get a computer to churn them out for me by just applying the rules of logical deduction from each previous statement. All of them would have an objective truth about them. All of them are mathematically true statements. But the point is that just as most of the books in the library of Babel are not interesting, so too most of these new theorems are banal or without interest.

There is more to mathematics that just generating mathematical truths. The art in being a mathematician is to single out those logical pathways that have something special about them. And here I think a sense of aesthetics plays a key part in making those choices. The reason I wanted to tell everyone about my discovery of this new group of symmetries is that it was surprising. It was like a moment in a novel when you think the main character is one thing but then suddenly he mutates into something quite different.

Maybe the difference between mathematics and the other sciences is that it is the natural world that is acting as the agent, picking out those things which have a special quality about them and we, as scientists, then try to understand why they are so special and selected. Often the answer is ultimately a mathematical one.

I like the suggestion made elsewhere in this volume that quite often mathematics is valued when you seem to get more out than you put in. The definition of a group of symmetries looks quite simple. It's hard to believe that it has led to the discovery of strange objects like the Monster and E_8.

Cultural and historical context have an effect on the reception and excitement over different mathematical discoveries. Every 21st century mathematician cares whether there is a zero of the Riemann zeta function sitting off the critical line. Even though it would be an impressive feat of mathematics, I just don't think 21st century mathematicians care if there are odd perfect numbers or not. That's why no one is really working hard to prove this fact. In contrast, the ancient Greeks might have got very excited by the discovery. Of course the proof might yield exciting new insights about numbers that mathematicians might value. Does anyone really care whether Fermat's equations $x^n + y^n = z^n$ have integer solutions or not? Certainly there aren't many theorems that started 'suppose Fermat's Last Theorem is true then . . . ' Why the mathematical community continued to pursue a proof of this theorem is because it was a catalyst for the discovery of some amazing ideas.

One might try to make the distinction between mathematical and artistic creations by declaring that mathematics is discovering eternal truths about the universe. I can't make a theorem true just because I think it will be beautiful. If the Riemann Hypothesis turns out to be false it will shatter our sense of how beautifully we believe the primes are laid out. But there will be nothing we can do about it. The Riemann Hypothesis is either true or false and there is no act of creative thinking which is going to alter that. In contrast, one can't talk about the objective truth of 'The Death and the Maiden' or *Great Expectations*. For a start, the works elicit multiple reactions from audiences. Ambiguity is an important part of creating art. Ambiguity for the mathematician is anathema. But the creative act involved in doing mathematics is the act of focusing on asking the question whether the Riemann Hypothesis is true. There are lots of questions that we can ask about the primes. Why this one is the Holy Grail is again because it says something very special about the primes. The connection between the primes and the zeros of the Riemann zeta function can't help but bowl you over when you read about it for the first time. It is such an extraordinary transformation.

Another key point about mathematical discoveries is how integrated they are across the subject. This integration is often important in how a piece of mathematics is valued. A mathematical discovery which seems isolated from the mathematical mainstream, however surprising or beautiful, will probably not receive the same sort of attention as mathematics that has connections with other bits of the subject. The fact that the Riemann Hypothesis is so interconnected with so many other bits of mathematics is one reason this mathematics is valued. It's like the Internet. More links and the higher your mathematical Google ranking.

Perhaps musical and literary creations can survive better in isolation, although quite often one can only truly appreciate these creations in relation to what has gone before.

The quest to prove the Riemann Hypothesis raises the interesting question of whether there is a difference between proving a conjecture and constructing new mathematical objects. Of course the creativity involved in constructing a proof that establishes whether the Riemann Hypothesis is true or not is matched by the creative act of constructing new non-Euclidean geometries. But there does feel like a difference in the process. It is a bit like being an explorer. Riemann pointed out a far distant mountain. Those trying to prove the Riemann Hypothesis are trying to find some route through the mathematical landscape to arrive at that mountain. Bolyai's discovery of non-Euclidean geometry is like the explorer coming across a new island in the middle of the ocean that has never been seen before.

What about the question of whether mathematical objects really exist? I certainly am a Platonist at heart. There are some things out there that are independent of our existence or act of imagining them. Prime numbers, simple groups, elliptic curves. It is not mathematicians who made these things. But then maybe I am getting back to the feeling that my group of symmetries is simply the articulation of a mathematical entity that was there all along. I think I have some sympathy with Kronecker's statement: 'God made the integers; all the rest is the work of man.' That's not to say that the Riemann Hypothesis is made true or false by the work of man or woman. It is either true or false that the primes are distributed as Riemann's Hypothesis predicts. But it's the decision to obsess about this question of mathematics and not some other hypothesis that is the work of man or woman. Again I think that the act of the mathematician is to tell particular stories about the integers and to single out those which are genuinely interesting and surprising to other mathematicians. I would argue that the aesthetic judgement that singles out great mathematics shares a lot in common with the traits that one is looking for in a great piece of music. It is very rarely the usefulness of a piece of mathematics that motivates a mathematician. It is often centuries later that a mathematical discovery ends up being applied to the real world. Rather, the mathematician is drawn to mathematics that is full of beauty, elegance and surprise. In a mathematical proof, themes are established then mutate, interweave, producing surprising moments of connection. These are traits that for me make exciting music as well as exciting mathematical arguments.

For a mathematician, the journey of that proof, either carving the pathway out for the first time or following in someone else's footsteps, is the essence of the mathematics, not the stark statement of the theorem being proved. For example, Fermat discovered the amazing fact that every prime number that has remainder 1 on division by 4 can always be written as two square numbers added together. For example 41 is a prime number which certainly has remainder 1 on division by 4. Fermat's theorem guarantees that this prime can be written as the sum of two squares, in this case: 25 + 16 or 5 squared plus 4 squared.

For a mathematician this theorem is exciting because it connects two very different sorts of numbers: primes and squares. But for the mathematician it is reading a proof of why there is this connection which provides the real pleasure. The kick comes from the moment when you suddenly see why there is a common thread connecting the squares and these primes. They are like two different variations on a common theme.

Just in the same way that people have started to quantify what makes good music by trying to plot its different characteristics, it might be possible to give some measure of why we regard some proven statements of mathematics as being worthy of prizes and publication in the *Annals of Mathematics* while others are just ignored as uninteresting. Is it to do with a certain complexity of the proof? Sometimes. Although simplicity is often a guiding light for a mathematician. The proof of the four-colour-map problem is complex but not beautiful and doesn't provide quite that magic 'ah-ha' moment when you suddenly get why four not five colours suffice. The proof of Fermat's Last Theorem is extraordinary and complex (and certainly was not what Fermat couldn't fit in the margin) but still a mathematician reading it is swept along by the twists and turns of ideas like a grand Wagnerian opera arriving at the final Q. E. D., realizing what an inevitable journey Wiles took you on. Another measure of the significance of the mathematics is related again to this idea of getting more out than you put in. A third measure is how integrated the result is with other mathematics that is valued—that mathematical Google rating.

But maybe trying to quantify what makes good mathematics is just as doomed to failure as trying to measure why Mozart's music is so magical. As Hardy writes in *A Mathematician's Apology*: 'It may be hard to *define* mathematical beauty, but that is just as true of beauty of any kind—we may not know quite what we mean by a beautiful poem, but that does not prevent us from recognizing one when we read it.'

I often feel that the create/discover question shares something with the nature/nurture debate. How much is a child just a product of their genes? Does the environment have much more influence on the outcome and characteristics of a child? The theorems that the mathematician discovers are their children, their legacy. The birth of a theorem is often preceded by long hard labours. Their existence is a way of continuing our legacy. The permanence of their proof is our chance of a bit of immortality. But are these theorems simply a consequence of the logical framework we work within, like some genetic code forcing their behaviour and existence? Or is our nurturing of those theorems we create a function of the culture, the surrounding environment of the mathematics that exists around us? It's not a very satisfying answer for a mathematician who likes things black or white, true or false, proved or disproved, but it's probably a little bit of both. But maybe that's why at the end of all these philosophical musings mathematicians so often just stick their heads back in the mathematical sands and continue trekking through its beautiful landscape, proving new theorems, constructing new mathematical structures, revelling in its unchanging certainty. This is the job of a mathematician.

Comment on Marcus du Sautoy's 'Exploring the mathematical library of Babel'

Mark Steiner

———————≫◦◦◦◦═———————

Professor du Sautoy reconciles the realist and the constructivist positions in the philosophy of mathematics with a simple, but effective distinction. Structures describable in mathematical language exist independently of our knowledge; this is the realist part. The mathematician chooses, from among these structures, those which are to be called mathematical structures. To be describable in mathematical language is not yet to be a mathematical structure. Professor du Sautoy adds that aesthetical considerations play a dominant role in deciding what is worth investigating, i.e., what is to be called mathematics. This is exactly the position I took in my book, *The Applicability of Mathematics as a Philosophical Problem* (1998). The way I put it is that mathematics is *anthropocentric* to the extent that anthropocentric criteria (like aesthetical) govern what is called mathematics.

What now of Hardy's view that beautiful mathematics is never 'useful', which du Sautoy quotes approvingly? I do not see any reason for Professor du Sautoy to accept what is a patently false view, based mainly on wishful thinking. (Hardy didn't want mathematics to be used for warfare.) Hardy is so spectacularly wrong that, on the contrary, many scientists are convinced that the more beautiful mathematics is, the more applications it has. Hardy wrote, 'No one has yet discovered any warlike purpose to be served by the theory of numbers or relativity, and it seems very unlikely that anyone will do so for many years.' While the view of some that Albert Einstein invented the atomic bomb is ludicrous, to say on the other hand that there is no warlike purpose for the equivalence of mass with energy is equally ludicrous. And as for number theory, much of the work in the field, I am told, is simply classified, because it could be used, and is used, in cryptography. If somebody came up with a good algorithm for factoring large numbers he would probably be arrested.

I leave Professor du Sautoy, with the following challenge: can you think of an explanation why beautiful mathematics tends to be useful in applications?

3
Mathematical reality
John Polkinghorne

Are mathematicians engaged in acts of discovery or are they merely constructing ingenious intellectual puzzles whose solutions simply afford occupation
and amusement for those whose tastes lie in that direction? Is mathematics just
the painstaking unravelling of a monstrous logical tautology? Or is mathematics something much more interesting and significant than either of these rather
banal judgements would suggest?

Seeking answers to these questions is not just a way of assessing the dignity
and importance of mathematics itself, for the result of the enquiry promises
also to provide a significant source of insight into the discussion of wider
and deeper philosophical issues. The status of mathematics bears upon an
answer to the fundamental metaphysical question, 'What are the dimensions
of reality?' Do they extend beyond the frontiers of a domain that is capable of
being fully described simply in terms of exchanges of energy between material
constituents, located within the arena of space-time? For the materialist, the
latter is indeed the true extent of reality, and all other human talk, such as
that employing mental or axiological categories, amounts to no more than
convenient manners of speaking about epiphenomena of the material. Or, on
the contrary, is it the case that true ontological adequacy requires that much
more be said than physicalism can articulate?

The issue of the nature of mathematical entities provides a convenient test
case for probing this general question. Particularly helpful as an introduction
to the considerations involved in pursuing the matter is the published report
of an extended conversation between two distinguished French savants, Jean-
Pierre Changeux, a molecular neurobiologist and a resolute materialist, and
Alain Connes, a mathematician and a firm believer in mathematical reality
(Changeux and Connes, 1995). Changeux asserts that mathematical entities
'exist in the neurons and synapses of the mathematician who produces them'
(ibid., p. 12), while Connes claims that he finds in the world of mathematics
'a more stable reality than the material reality that surrounds us' (ibid.). Two
radically different metaphysical positions stand opposed to each other in this
confrontation.

Metaphysics

One of the fundamental questions in philosophy is how we should conceive that epistemology (knowledge) and ontology (being) are related to each other. At one extreme, someone like Immanuel Kant divorced the two. In his opinion, all we can know are phenomena, the appearances of things, while noumena, things in themselves, are hidden from our view. At the other extreme stands the realist, for whom epistemology and ontology are closely correlated and what we know should be taken as a reliable guide to what is the case. Almost all scientists, wittingly or unwittingly, are realists. If they did not believe that scientific knowledge is telling us what the world is actually like, it is difficult to see what would justify the great expenditure of labour and mental energy involved in pure scientific research. Even in science, however, physics constrains but does not fully determine metaphysics. There is no simple entailment between the two. Their mutual relationship is similar to that between the foundations of a house and the variety of edifices that might eventually be erected upon them. For example, quantum physics is undoubtedly intrinsically probabilistic, but does this uncertainty arise from a necessary ignorance of deterministic fine detail or is it the sign of an actual intrinsic indeterminacy present in nature? While most physicists follow Niels Bohr and his successors in taking the latter view, David Bohm showed that there is an alternative interpretation of the theory that yields identical physical predictions but which corresponds to the former option of a veiled determinism (Bohm and Hiley, 1993). Since there is no difference in the empirical consequences involved between the two, the choice between Bohr and Bohm cannot be made on strictly physical grounds but appeal must be made to metaphysical considerations, such as naturalness and lack of contrivance.

Rather similar problems arise in relation to the nature of mathematical entities. There is a vast and impressive body of mathematical knowledge. What, if any, kind of reality is accessed by this knowledge? In the course of defending his position, Changeux says that he adopts 'a naturalist position that makes no reference whatsoever to metaphysical assumptions' (Changeux and Connes, 1995, p. 213). How difficult it is to see ourselves as others see us! Anyone who outlines a world view circumscribing the scope of reality, whether it is narrowly or capaciously conceived, is making a metaphysical assertion as surely as they are using prose to express their convictions. The materialist is no exception to this rule. It is a common illusion, often entertained by those taking a reductionally physicalist position, to suppose that somehow they are exempt from having to make prior metaphysical assumptions, other than a reliance on the reliability of science. Yet the fact is that science has purchased its very great success by the limited character of its actual aims. We have seen from the example of quantum physics that its discoveries constrain metaphysical thinking, but they are by no means sufficient in themselves to determine what its conclusions should be. The reduction of thought solely to physical states of neural networks is not a deduction from neurobiology, but a metaphysical assumption imposed upon that scientific discipline. Of course, I do not challenge the belief that there is a connection between the workings of the mind and the behaviour of the brain—I certainly accept that human beings are psychosomatic entities—but the nature of that relationship is a not a question that can be settled by neurological investigation alone, however important and interesting that investigation undoubtedly is. Metaphysical questions demand metaphysical answers, which have to be supported by metaphysical arguments.

Contemporary society, in striking contrast to the thinking of many previous ages, seems to treat materialism as the natural default position, scarcely requiring any argument in its defence. Yet the world picture it presents is that of a kind of lunar landscape, with complex, replicating and information processing systems as its inhabitants, but with no persons in it. Much of what makes human life valuable and satisfying is dismissed as epiphenomenal. No due acknowledgement is given of the creative powers of imagination involved in the intellectual enquiry that gave birth to science and mathematics. Personal experience, which is the foundation of all our most significant encounters with reality, rather than being accorded the privilege it deserves, is dismissed with an unwarranted suspicion of its importance. Our mental life—the actual source of all our knowledge—is treated as if it were a by-product of the material, in a curious replacement of the direct by the abstracted.

In the materialist perspective, human beings are simply seen as computers made of meat. The unsatisfactoriness of this as far as mathematical experience is concerned seems clear. Mathematical thinking is more than computational efficiency; mathematical insight is not confined within the Gödelian limits of finitely axiomatized systems, a point that has been emphasized particularly by Roger Penrose (Penrose, 1989).

Mathematical reality

Considerations relating to the issue of the reality of a noetic world of mathematical entities bear some analogy to similar arguments that can be made defending the reality of the physical world against the critiques of the idealists. However, before going on to consider these metaphysical arguments, one must first be clear what results might be expected of them. The character of the conclusions reached will be insightful and persuasive, rather than logically coercive. The strict language of 'proof', with the implication that only a fool could disagree, is inappropriate in this field of discourse. No one can force an intransigent sceptic to give up their position, however arid and implausible it may be. The solipsist, and the person who maintains that the world and our memories of it came into being five minutes ago, are both logically invulnerable in their absurdities. The best that can emerge from metaphysical disputation is an argued claim to have attained the best explanation that is available.

The first of the analogies between human encounters with the physical world and with the mathematical world relates to the consistency of perception, and the mutual coherence of account, reported by different observers. Connes summarizes this argument by saying,

> *What proves [too strong a word!] the reality of the material world, apart from our brain's perception of it? Chiefly the coherence of our perceptions and their permanence. . . . And so it is with mathematical reality: a calculation carried out in several different ways gives the same result, whether it is done by one person or by several.*
>
> Changeux and Connes (1995, p. 22).

A second argument appeals to the richness to be found in independent reality. Science's exploration of the physical universe is the story of the discovery of an apparently unending depth of rational structure and relationality, unveiled as level after level of the world is disclosed to human enquiry. The richness thus revealed is strongly persuasive that its source lies outside the limited human mind of the investigator. For mathematics, Connes invokes Gödel's theorem, with its implication that the richness of arithmetic will never be contained within a system of finite axiomatization, to make an analogous point about mathematical reality. He says that the theorem means that 'the quantity of information contained in the set of all true propositions about the positive integers is infinite', going on to comment 'I ask you: isn't that the distinguishing feature of a reality independent of all human creation?' (ibid., p. 160).

A third argument, related to the last, points to the element of surprise involved in the exploration of an independent reality. A powerful support for a realist interpretation of physical science can be found in the manner in which the universe frequently resists prior expectations about its character, forcing upon the physicist concepts that would never have been accessed without the relentless pressure exerted by the stubborn nudge of nature. The resulting feel of the scientific endeavour is that of discovery rather than construction. The counterintuitive ideas of quantum physics are perhaps the most striking example of this phenomenon. Who would have supposed that the apparent ambiguity of wave/particle duality was a rational possibility without being driven to it by the stubborn facts of the observed character of light? In a somewhat similar way, totally unanticipated riches are revealed to the explorers of the mathematical world. Connes' favourite example is provided by the 26 'sporadic' finite simple groups, which defy classificatory incorporation into such general categories as cyclic groups of prime order (du Sautoy, 2008). A more pictorially accessible example would be the endlessly proliferating structure of the Mandelbrot set, deriving from a deceptively simple-looking and concise definition.

Considerations of this kind help to explain the conviction held by many mathematicians that they are engaged in the discovery of actually existing entities and their properties, and not merely the invention of pleasing intellectual games, indulged in simply to exhibit their skill. In his book, *A Mathematician's Apology*, the distinguished mathematical analyst G. H. Hardy stated his conviction that 'mathematical reality lies outside of us, that our function is to discover or observe it, and that the theorems which we describe grandiloquently as our "creations", are simply our notes of our observations' (Hardy, 1940: 1967, pp. 123–124). Of course, our material brains are involved as instruments in making these observations, just as they are in our making observations of the physical world around us, but in neither case should the means of perception be equated with the realities perceived. Changeux's attempt to reduce mathematical entities to items of synaptic storage is to be resisted as a category mistake, as crass as identifying literature with the ink and paper by means of which it is recorded.

The plausibility of the concept that mathematical research is an act of noetic exploration is strengthened by the role played in mathematical thinking by intuitive perception and unconscious creative activity. Something is going on that seems much more profound than can be described by a banal concept of computational processing. There are well-documented cases of discovery in which, after intense conscious engagement has failed to yield the solution of a deep problem, a period of fallow disengagement is followed by a moment of illumination in which the answer emerges into consciousness, essentially fully-formed and only needing extended technical labour to complete the details of proof. A well-known case involved the nineteenth-century mathematician, Henri Poincaré. He had been wrestling with a problem connected with the theory of Fuchsian functions, but he had made no progress. Consequently, Poincaré decided to give it a rest and to take a holiday. At the very moment of his departure, the complete solution to the problem sprang into his mind unbidden. So sure was he that he had made the breakthrough, that he continued on holiday, only engaging with the technical mopping-up operation on his return. Perhaps the most striking example of the existence of profound intuitive mathematical powers was given by Hardy's Indian colleague, Srinivasa Ramanujan. This self-taught genius displayed an astonishing ability to write down deep theorems in number theory that he had discovered, not by explicit rational argument but by a tacit process of intuitive encounter. It is surely more persuasive to understand Ramanujan's great gifts as the consequence of an ability to access and explore an existing noetic world, rather than there being simply some fortuitous tricks of his neural organisation.

Evolution

A final argument for taking the independent reality of mathematical entities seriously derives from asking how it might have been that profound mathematical ability arose in the course of hominid evolution. It seems clear enough that some very modest degree of elementary mathematical understanding—the ability to count, simple notions of Euclidean geometry, and the capacity to make simple logical associations—would have provided our ancestors with valuable evolutionary advantage. But whence has come the human capacity to go far beyond matters of everyday utility, to attain the ability to conjecture and eventually prove Fermat's Last Theorem, or to discover non-commutative geometry? Not only do these powers appear to convey no direct survival advantage, but they also seem vastly to exceed anything that might plausibly be considered a fortunate spin-off from such mundane necessity.

The power of evolutionary explanation depends critically on getting the environmental factors right, as much as it does on getting the genetic factors right. If the context within which hominid evolution took place was solely that of the physico-biological dimensions of reality, as strict neo-Darwinian orthodoxy supposes, the coming-to-be of human mathematical ability would seem to be an inexplicable excess. Yet one can take Darwinian explanation absolutely seriously without having to suppose it to be a totally adequate account of absolutely everything that has happened. If mathematical entities constitute an independent realm of reality, then mathematics has always been 'there', even before mathematicians emerged. It formed the noetic context within which that emergence eventually took place. While survival pressures would favour the initial development of a brain structure that afforded access to limited arithmetical and geometrical thinking, once that modest degree of contact had been established with mathematical reality, further new developmental factors would come into play. The drive to assist physical survival would be supplemented by the effects of a mental influence that one may call 'satisfaction' (Polkinghorne, 2005, pp. 54–55). Intellectual delight would then draw our ancestors into an exploration of the noetic world of mathematical entities, beguiling them to progress far beyond the modest needs of everyday practicality. Doubtless the development of mental perceptive power that this involved was made possible by the epigenetic plasticity of the human brain, much of whose complex structure derives not from genetic inheritance, but from response to the shaping influence of experience. Belief in the reality of mathematics makes intelligible our human ability to be mathematicians, a capacity that otherwise would have seemed inexplicably gratuitous.

Unreasonable effectiveness

If mathematical entities are a part of reality, then one might expect that the ontological realm of their existence is not an isolated domain, disconnected from all else, but that it has subtle connections with other dimensions of the real. A very striking example of this happening is provided by the connection found to exist between theoretical understanding in physics and mathematical properties. It is an actual technique for discovery in fundamental physics to seek theories that are formulated in terms of equations possessing the unmistakable character of mathematical beauty. This beauty is a rather rarefied form of aesthetic experience, but it is one that mathematicians can readily recognize and agree about. It involves qualities such as elegance and economy and the property of being 'deep'; that is to say, extensive and surprisingly fruitful consequences are found to derive from an apparently simple starting point. The physicists' search for beautiful equations is no mere aesthetic indulgence but a heuristic strategy which has proved its worth time and again in the three-century history of modern physics. Paul Dirac, one of the founding figures of quantum theory, made his remarkable discoveries through a lifelong and highly successful pursuit of mathematical beauty. He once said that it is more important to have beauty in your equations than to have them fit experiment! Of course, Dirac did not mean that empirical adequacy was ultimately dispensable. No scientist could think that. If you have solved the equations of your new theory and found that the answers do not appear to agree with experiment, that is undoubtedly a setback. However, it is not necessarily absolutely fatal. No doubt you have had to have recourse to some approximation scheme in getting your solution, and maybe you have just made an inappropriate approximation. Or maybe the experiments were wrong—we have known that happen more than once in physics. So there would still be at least a residue of hope. But if your equations were ugly... well, there was no hope. The whole history of physics was against you.

Dirac's brother-in-law, Eugene Wigner, who also won a Nobel Prize for physics, once called this remarkable ability of mathematical beauty to unlock the secrets of the physical universe its 'unreasonable effectiveness'. How does it come about that this apparently abstract subject can illuminate our understanding of the structure of the physical world? Why are the beautiful patterns of pure mathematics, discovered by the mathematicians in their studies, so often found actually to occur in the structure of the world about us? This is not the place to pursue that particular issue in detail (I personally look to natural theology for an answer (Polkinghorne, 1998, ch. 1)). It is sufficient for our present purpose simply to note the fact, and its implication of a deep mutual entanglement of the physical and the mathematical. Few doubt the reality of the physical world; they should be prepared to consider acknowledging a similar reality of the mathematical world that intertwines with it.

Mathematics also entangles with other dimensions of reality. I wish to take very seriously human encounter with the realm of beauty. I do not think that our aesthetic experiences are simply some kind of epiphenomenal froth on top of what basically is just a physical substrate, but they are a form of access to yet another dimension of reality. Of course, music involves vibrations in the air, but its appreciation is not to be reduced to the fourier analysis of those vibrations. There is a deep mystery about the way that the impact of packets of sound waves on the eardrum can evoke in us what I believe to be the valid experience of encounter with a timeless beauty. There is an often recognised kinship between mathematics and music, expressed not only by the way that individuals frequently display interests and skills in both, but also by the patterns that both are found to share, particularly in the case of contrapuntal music.

Another aesthetic experience in which patterns play a vital is exhibited to us by Islamic art. Marcus du Sautoy (2008, ch.3) has given a fascinating discussion of the symmetries that underlie the elaborate decoration of the walls of the Alhambra Palace in Granada. In the nineteenth century, group theorists were able to show that there are just 17 different basic kinds of symmetry that can be present in such regular coverings of the plane. All 17 are present in the decorations of the Alhambra, which was built in the thirteenth and fourteenth centuries. The Islamic artists involved did not know the group theory, but they had intuitive access to the mathematical reality that their work expressed.

Conclusion

The criterion for assessing the persuasiveness of a metaphysical position is the seriousness with which it treats, and the adequacy with which it can contain, the great swathe of basic human experience which it is seeking to make intelligible through its insight. Schemes that are made parsimonious simply by the illegitimate Procrustean strategy of the excision of what, from their point of view, is inconvenient to take into account, are certainly to be rejected. The argument of this chapter has sought to show that an approach that seeks to take the actual character and achievements of mathematics with the due seriousness that they deserve, is one that is best formulated in a metaphysical context that acknowledges the reality of a noetic world of mathematical entities.

Comment on John Polkinghorne's 'Mathematical reality'

Mary Leng

———=➤∞➤=———

Attacking the thorny issue of 'creation vs. discovery' in mathematics, John Polkinghorne's contribution to this volume argues that mathematicians are engaged in inquiry into the nature of a 'noetic realm' of mathematical entities, and therefore in discovery, rather than creation, of their mathematical subject matter. Polkinghorne contrasts this picture with a physicalist alternative, according to which human mathematical activity would have to be explained without reference to such a realm of mathematical objects, for example, as the construction of 'ingenious intellectual puzzles' or 'the painstaking unravelling of a monstrous logical tautology'.

Polkinghorne quite rightly notes that no deductive proof can be provided for his, or indeed any, account of the fundamental nature of mathematics: our philosophical theory here is underdetermined by the empirical evidence, so at best *inductive* reasons (reasons which do not establish, but at best make probable, their conclusion) can be given to prefer Polkinghorne's noetic realm hypothesis to the hypothesis of physicalism. Polkinghorne further notes that physicalism should be viewed as just one metaphysical hypothesis among others, rather than as a default position to be held on to unless one is provided with conclusive reasons for its rejection. Polkinghorne's strategy is therefore to consider the noetic realm hypothesis alongside physicalism's denial of this realm, subjecting both to the evidence provided by a variety of phenomena, and to argue that these phenomena speak against physicalism and in favour of the existence of a noetic realm of mathematical objects.

That a question is only amenable to inductive considerations, and not deductive proof, does not of course take away from the meaningfulness, value, importance, or indeed tractability of that question. Indeed, while pure mathematics thrives on deduction, inductive reasoning is the lifeblood of the empirical sciences, where almost every important theoretical question requires a leap beyond what can be narrowly deduced from empirical observation.

Polkinghorne's discussion makes use of two distinctive types of inductive reasoning that are commonplace in empirical science in order to argue for his metaphysical conclusion: arguments by analogy, and inferences to the best explanation.

Arguments by analogy take the general form: X is like Y in respects $a, b, \ldots$. Therefore (probably), X is also like Y in respect z. This form of reasoning is essential, for example, in drawing conclusions about real situations from models of those situations, where the inferences can be quite mundane. (The city is like our map with respect to the placement of the landmarks, streets, and relative distances that we can see. Therefore (probably), it is also like our map with respect to the location of the town hall. So to get to the town hall we should go straight ahead and take the third street on the left.) More creatively, Newton made use of an analogy between planets and balls to argue that the motion of a planet around the Sun should be relevantly similar to the motion of a ball being swung around on an elastic string (with the circular motion being produced by the 'pulling' force exerted by the elastic in the ball case, and by the Sun in the planet case). Speculative analogies of this sort require some crafting. For any values of X and Y, we will be able to find some respects in which they are similar. The trick is to find enough *relevant* similarities to warrant the inference to a further similarity.

Polkinghorne's argument by analogy (which he draws from Alain Connes's response to Jean-Pierre Changeux's radical materialist position) makes use of three respects in which mathematical inquiry is like inquiry into the material world. His conclusion is that mathematical inquiry is also like inquiry into the material world in having an independent, objective, real world of objects as its subject matter. Polkinghorne himself suggests that each of these respects is, by itself, reason enough to draw his realist conclusion, but taken together, they can only strengthen the overall analogy. In assessing this analogy, then, it is important to consider whether the respects in which the two forms of inquiry are similar (coherence of 'perceptions' across time and observers; richness; and capacity to surprise) are relevant in relation to the further similarity Polkinghorne wishes to claim (i.e., concerning an independent realm of objects). Polkinghorne has certainly picked on markers of *objectivity* in mathematical reasoning: that different reasoners using different methods in isolation from one another can still agree on mathematical conclusions, for example, certainly suggests that mathematicians are not free to draw whatever conclusions they please. But are the similarities relevant to Polkinghorne's further claim, that mathematical inquiry, like physical inquiry, concerns an objective, independent, realm of *objects*?

In answering this question, one might consider whether an alternative account of the nature of mathematics can explain the phenomena Polkinghorne mentions as well as, or better than, does Polkinghorne's noetic realm hypothesis. If some alternative, and plausible, explanation of the similarities between mathematical and physical inquiry can be provided, which does not appeal to a noetic realm, then the strength of the original argument from analogy is undermined. This brings us, then, to the second of Polkinghorne's argumentative strategies: inference to the best explanation (IBE).

Our late colleague, Peter Lipton, describes this form of inference as follows:

> *Given our data and our background beliefs, we infer what would, if true, provide the best of the competing explanations we can generate of those data (so long as the best is good enough for us to make any inference at all).*

> *Lipton (1991: 2004, p. 56)*

As Lipton points out, the word 'best' needs some clarification here. In particular, we can distinguish between

> *the explanation best supported by the evidence, and the explanation that would provide the most understanding or, in short, between the likeliest and the loveliest explanation.*

> *Lipton (1991: 2004, p. 59)*

Advocating inference to the likeliest explanation is, as Lipton points out, relatively uncontroversial but, sadly, fairly unhelpful—if we had a way of knowing what scenario was most *probable*, we would certainly infer that over the alternatives, but IBE is surely intended in part as a method for discovering which of alternative possibilities is more probable. For inference to the best explanation to be a practical rule of theory choice, we need to build on our account of what makes an explanation *lovely* in the sense of providing understanding—perhaps by invoking theoretical virtues such as simplicity, non-ad-hocness, unifying power, and so on. Whatever our account of 'loveliest' amounts to, Polkinghorne clearly thinks that the noetic realm hypothesis provides the loveliest explanation of the similarities between mathematical and physical inquiry he indicates. Lovelier, certainly, than Changeux's account of mathematical objects as existing 'in the neurons and synapses of the mathematician who produces them'—could those neurons and synapses really contain the rich, surprising, and universally accessible subject matter of mathematical inquiry? And there are further phenomena for which Polkinghorne thinks the noetic realm also provides the loveliest explanation. In particular, Polkinghorne points to cases of sudden and deep mathematical insights (such as those reported by Poincaré and Ramanujan); the capacity of human reasoners to go beyond the narrow range of mathematics that we could expect to give us evolutionary advantage; and the 'unreasonable effectiveness' of mathematics in empirical science. To take just the last of these, Polkinghorne argues that seeing mathematics as one dimension of reality renders its effectiveness in finding out about physical reality unsurprising, since one should expect the mathematical realm to 'have subtle connections with other dimensions of the real'. But that two systems of objects *exist* cannot by itself be enough to explain why facts about one system are relevant in finding out about facts about the other: my kitchen exists, and so does the solar system, but if it turned out I could reliably divine facts about the solar system by reasoning about the contents of my kitchen, one might still find the effectiveness of this reasoning unreasonable. More needs to be said about why the mere *existence* of a mathematical realm should render its effectiveness reasonable.

There are those who look with suspicion on metaphysical speculation, on the grounds that it is often unclear what kind of evidence could count in favour of, or against, any particular metaphysical hypothesis. One of the great merits of Polkinghorne's contribution to this volume is that it lays out in clear and precise terms what would be required of his opponent in this debate: find a better explanation of the phenomena in question that does not make use of the noetic realm hypothesis. For those (myself included) who think alternative explanations can be found, Polkinghorne's paper presents a formidable challenge.

Reply to Mary Leng

John Polkinghorne

———————

I am grateful to Mary Leng for her helpful analysis of the arguments I sought to deploy in defense of mathematical realism. Of course she is right that the existence of two objects does not itself imply a mutual connection, but the relationship between mathematics and physics is a deep and apparently intrinsic (unlike her table and the solar system), and I still maintain this encourages the thought that each is a part of a greater reality.

4
Mathematics, the mind, and the physical world

Roger Penrose

Does mathematics have an independent reality? Or is it simply a product of human thought and culture? Or perhaps it is merely an idealization, abstracted from what we find to be a mathematical organization that is well approximated in the structure and dynamical behaviour of the physical world?

In this chapter I shall attempt to address these two aspects of mathematical Platonism: the issue of the independent reality of mathematics and the separate issue of whether there is a fundamental dependence of physical behaviour on such a pre-existing mathematics. These two themes, which perhaps are often confused with one another in the minds of those who are uncomfortable with mathematical Platonism, form the essential subjects of this chapter. I shall try to illustrate my own position on these issues in relation to Fig. 4.1, where I have schematically depicted[1] three 'worlds': the physical, the mental and the mathematical, together with what I regard as the three deeply mysterious connections between them.

The first aspect of mathematical Platonism referred to above is the nature of 'Mystery 0', namely whether the 'world of mathematics' arises merely as a product of our mental activities, having no reality beyond this, or whether it is to be assigned an independent existence of its own. And if the latter, whether or not we, in principle, have access to this world in its entirety. The second, separate, issue—depicted as 'Mystery 1'—has to do with the role of mathematics in physical theory. Does the undoubted utility of mathematics in our understanding of the physical world reflect merely our facility in organizing observational data into some comprehensible form, where those aspects of physics which work well as mathematical theories are assigned an undue importance, merely because they *do* work well? Or is it really true, as

---

[1] This figure first appeared in Penrose (1994), but I have used it frequently elsewhere, such as in Penrose (2004).

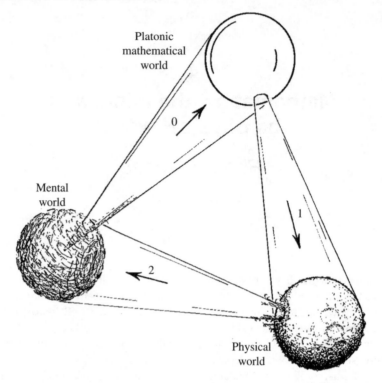

Platonic
mathematical
world

0

Mental
world

1

2

Physical
world

Fig. 4.1 The three 'worlds': the physical, the mental and the mathematical.

many theoretical physicists appear to believe, that there is a deep and precise underlying dependence of the operation of the physical world on a pre-existing mathematical order—a mathematical order that appears to have great beauty and sophistication—which is there to be discovered and not simply imposed upon Nature as a feature of our gropings towards understanding?

To complete the triad of mysteries connecting these worlds is 'Mystery 2', which is concerned with the relation between physical reality and mentality, particularly *conscious* mentality. How does consciousness arise in a world which seems to be governed by entirely impersonal mathematical operations? Or is consciousness in some sense primary, its presence being an essential prerequisite for the very existence of a structure that we could, in any sense, call a 'universe'? Is it merely the complication, or perhaps some other sophisticated quality, in the construction of our brains that allows this mysterious phenomenon of consciousness to come about? And, if so, is this complication or sophistication to be understood solely in *computational* terms, as seems to be a common viewpoint in our present computer age. Or is there some other essential prerequisite to consciousness, that cannot be understood in purely computational terms? If the latter, is this ingredient something lying hidden in the physics that we presently use for our descriptions of the operations of the world? Or must we search for deeper (mathematical?) theories for a physical description of consciousness to be possible. Or might we perhaps have to look even farther afield, to an understanding that lies essentially beyond any kind of science whatever, as could be the implication of an essentially religious perspective on these issues?

In this chapter I shall be concerned mainly with Mysteries 0 and 1, as the topic of the Symposium had to do specifically with mathematics. But it is my opinion that an adequate discussion of these two mysteries cannot be completely divorced from some discussion of Mystery 2. I shall argue the case (by appealing to Gödel incompleteness) that the very fact that our minds are capable of comprehending sophisticated mathematical arguments—at least under favourable circumstances—leads us to the conclusion that the operation of conscious minds cannot be entirely computational and, accordingly, that our minds cannot be the product of an entirely computational physics. It does not appear to be the case that the physical laws that we presently understand contain anything that is essentially non-computational (where merely 'random' is not to be considered as 'essentially non-computational'). The conclusion from this is that there must be something beyond our present-day physical laws that is operative in the actions of a conscious mind. It is my own viewpoint that this is strongly indicative of the actions of a conscious brain being dependent upon areas of physics (probably at the quantum/classical boundary) that lie outside the scope of our present-day physical theories; yet that the needed physical revolution may not itself lie too far beyond what is presently understood.

The issue of Mystery 0 is indeed a closely related matter. Part of the reason for regarding our access to mathematical truth as being something 'mysterious' lies in the nature of our ability to perceive the truth of various particular mathematical assertions. As Gödel (and Turing) have demonstrated, if we accept any particular computationally checkable system of procedures P as providing valid methods of mathematical proof, then we must equally accept the truth of some proposition $G(P)$, where the truth of $G(P)$ lies beyond the scope of the procedures P. Consequently, our methods of ascertaining mathematical truth cannot be entirely reduced to computational procedures that we accept as valid. Although various logicians have attached different interpretations to this conclusion, in my own view, it has the clear implication that there must be something essential lying outside pure computation that is operative in conscious understanding. (For further discussions of this point, see Penrose, 1997.) But what *really* is going on in the activity of a conscious mind when it becomes convinced of the truth of some mathematical proposition remains profoundly mysterious.

I maintain, also, that this is strongly indicative of mathematical truth being something objective (as was, indeed, Gödel's own view), and is not merely some 'game' based on arbitrary rules that arise out of human culture. Yet, I am prepared to accept that there could be 'degrees of Platonism', where some mathematicians might regard the truth of some proposition P as being an 'objective' matter, whereas another might take the view that the 'truth' or 'falsity' of P is a matter of opinion, depending upon what 'man-made' axiom system is adopted. I believe that it is clear, from Gödel's construction, that certain areas of mathematics are indeed 'objective', and therefore have an existence that is outside ourselves. Such an area would be the truth of what are called 'Π_1-sentences', these being assertions of the form 'such-and-such a computational procedure never terminates' (where 'computational procedure' means 'Turing-machine action'). An example of a Π_1-sentence is the famous Fermat's Last Theorem. It seems to me that the truth or falsity of a Π_1-sentence is something entirely objective, and so the truth values of Π_1-sentences have a Platonic reality which is not in doubt (although there may be an element of subjectivity to the issue of whether or not some particular Π_1-sentence has actually been established).

On the other hand, the objectivity of more sophisticated assertions such as the truth or otherwise of Cantor's (generalized) continuum hypothesis is perhaps more questionable, and it would seem to require a stronger form of mathematical Platonism to require that all such assertions are true or false in an absolute sense,[2] and not merely dependent upon some particular 'man-made' axiom system. The issue of whether or not some particular mathematician is, or is not, a Platonist usually refers to Platonism in this stronger sense. My own position is not to be particularly troubled by this issue, as a relatively weak form of Platonism seems to be perfectly adequate for most arguments of relevance to physics.

Accepting that, indeed, we need to consider a 'Platonic mathematical world' that is merely large enough to encompass the description of physical laws, we have an additional case for its 'existence', lying beyond anything that is merely brought into being by human culture or imagination. For the operations of the physical world are now known to be in accord with elegant mathematical theory to an enormous precision. One particularly striking example (see, for example, Hartle, 2003) is the double-neutron-star system PSR 1913+16, which has been under observation, now, for some 30 years, and the agreement between observation of the timing of its pulsar signals and Einstein's general theory of relativity (to something like one hundredth of a thousandth of a second over that entire period) is phenomenal. This indicates an extraordinary concurrence between the workings of the natural world at its most fundamental levels (here the very structure of space and time) and sophisticated mathematical theory. It makes no sense to me that this concurrence is merely the result of our trying to fit the observational facts

---

[2] It should be made clear that the results of Gödel and Cohen, showing that the continuum hypothesis is independent of the standard axioms of set theory, do not in themselves answer the question of whether or not this hypothesis is true in some absolute sense; see Cohen (1966).

into some organizational scheme that we can comprehend; the concurrence between Nature and sophisticated beautiful mathematics is something that is "out there" and has been so since times far earlier than the dawn of humanity, or of any other conscious entities that could have inhabited the universe as we know it.[3]

Acknowledgement

I am grateful to NSF for support under grant PHY 00-90091.

---

[3] More details about the arguments given here are to be found in Penrose (2011).

Comment on Roger Penrose's 'Mathematics, the mind, and the physical world'
Gödel's theorems and Platonism

Michael Detlefsen

———◦◦◦◦———

Roger Penrose's chapter contains a number of claims and ideas that warrant, and have received, extensive discussion. In this note, I will focus on the following two claims that are central to his view of the significance of Gödel's theorems.

I: Gödel's incompleteness theorem(s) demonstrate that 'if we accept any particular computationally checkable system of procedures P as providing valid methods of mathematical proof, then we must equally accept the truth of some proposition $G(P)$, where the truth of $G(P)$ lies beyond the scope of the procedures of P.'

II: Claim **I** 'has the clear implication that there must be something essential lying outside pure computation that is operative in conscious understanding.'

We can restate Penrose's claims more clearly and in more familiar terminology as follows:

I*: For any formal system P, if we accept all of P's axioms as true, and all its rules of inference as valid, we are rationally obligated to accept P's Gödel sentence $G(P)$ (and its P-equivalent consistency formulae $Con(P)$) as true too.

II*: Claim **I*** clearly implies that there is a set $\mathcal{A}$ of sentences we are rationally obliged to accept that cannot be formalized (i.e., is not a computably enumerable set).

There is little reason I can see for confidence in either **I*** or **II***. $G(P)$ and the usual consistency formulae $Con(P)$ equivalent in P to it are not logically implied by P. This means that they logically imply sentences that are not logically implied by P. There is generally speaking no reason to believe that all reasoning which supports P will equally support these 'extra' implications. Consequently, there is no general reason to think that rational acceptance of P will rationally oblige acceptance of $G(P)/Con(P)$.

This may seem wrong to those who believe that evidence capable of justifying P must also be capable of justifying belief in its consistency. It is important to realize, however, that to justify the consistency of P and to justify $G(P)$ or $Con(P)$ are not the same thing. Justification that P is consistent is not in and of itself justification of $G(P)$ or $Con(P)$. For the latter, justification for the following supplementary proposition is also required.

Supp: If P is consistent, then $G(P)/Con(P)$.

Such justification will not, however, necessarily be included in the evidence one might have for P's consistency.

This is not to deny, of course, that there may be justification for **Supp**. Nor is it to deny that there may be rationally compelling grounds for P that include rationally compelling grounds for $G(P)/Con(P)$. It is intended only to indicate the non-triviality of **Supp** and thus to demur to claims such as I^*, which suggest that rational acceptance of $G(P)/Con(P)$ is somehow presupposed by rational acceptance of P.

There are other reasons too to question Penrose's argument, but there is no space to enter into these here.

5
Mathematical understanding
Peter Lipton

I am alas not a philosopher of mathematics, but I am interested in the nature of scientific explanation and I am hoping that we can have a discussion of the nature of mathematical understanding, a topic on which some other members of the Symposium are expert. It would be enlightening for me if we could compare physical and mathematical explanation: the comments below flag some natural focal points.

My own work on explanation is motivated in part by two simple and general considerations. The first is that there is a gap between knowing that some phenomenon occurs and understanding why it occurs. Knowing is generally necessary but not sufficient for understanding. Thus most people know that the same side of the Moon always faces the Earth, but few people understand why. (The question begins to nag once you realize that the phenomenon requires that the Moon's period of orbit around the Earth be exactly the same as its period of spin.) Much of the work on the philosophy of explanation can be seen as attempts to answer the question of what bridges this gap between knowing that and understanding why. And the existence of the gap places a helpful constraint on adequate answers to the question of how to make the connection, since it shows that any model of understanding that treats what is necessary for mere knowledge as sufficient for understanding will be inadequate. Thus the Hempelian idea that a good explanation provides understanding of the phenomenon in question by providing reasons to believe that the phenomenon was to be expected is inadequate, since reasons for belief are in these cases required simply to know that the phenomenon occurs.

The second consideration that motivates me places an additional constraint on an acceptable answer to the gap question. This is the 'why-regress', a feature of explanation that many of us discovered as young children, to our parents' consternation. Given an acceptable answer to a 'why-question', it is almost always possible to ask why about the answer itself. In my view, the moral of this regress is not that explanation is impossible but rather that, in a manner of speaking, we can explain with what we don't understand. B may explain A and hence provide understanding of why A is the case even if B is not itself explained. Thus the fact that my student's computer crashed explains why her essay was late, even if nobody knows why the computer crashed. What this shows is that understanding is not some kind of 'super-knowledge' that is transmitted from explanation to phenomenon explained. Understanding seems not to be some special epistemic state, but rather additional information, information that need not itself have a special epistemic status.

These two considerations—the gap between knowledge and understanding and the why-regress—are uncontroversial in the context of physical explanation; but perhaps not in mathematics. Thus one could deny that there is any distinction between merely knowing that some mathematical statement is true and understanding why it is true. This would be a close cousin of the denial that there is a distinction between explanatory and non-explanatory proofs. And in the mathematical case one could also resist the moral I drew from the why regress, that the explanation need not have a special epistemic status. Thus one might hold that mathematical understanding flows from the special epistemic status of mathematical axioms, and that these axioms are where the regress stops. Is mathematical explanation different from physical explanation with respect to the gap and the regress?

Two of the most popular accounts of physical explanation are the causal model and the necessity model. According to the former, to explain is to provide information about the causal history of the phenomenon in question; according to the latter it is to show that the phenomenon was in some sense necessary—it had to happen. These two models seem strikingly inappropriate for mathematical explanation. If one opted for the causal model, it looks like there would be no mathematical explanations, since mathematical facts, however one construes these, do not appear to enter into causal relations. So no proofs would be explanatory. And if one opted for the necessity model, it looks as though every proof would be explanatory. Indeed it looks as though the contrast between knowing and understanding would disappear. For even if one knows a mathematical truth not on the basis of any proof but rather on the basis of expert testimony, one would still know that what one was told is necessary, since we know in advance that, for any mathematical statement, if it is true then it is necessarily true.

Fortunately, there are other models of explanation in stock. The obvious one to reach for is some version of a unification model, according to which we understand why something is the case when we see how it fits into a unified pattern. This vague thought could be articulated in various ways, but at least it seems to be a source or a criterion that might differentiate explanatory from unexplanatory proofs. But a unification model may well leave out some sources of mathematical understanding. Proofs by *reductio* might be good test cases. It is natural to look here for example of non-explanatory proofs, but it is not clear that these proofs fail to provide understanding (if indeed they do fail) because they fail to unify. It may be more natural to say that they fail because they do not show what 'makes' the theorem true, where 'making' is a form of determination of which physical causation is only one species. So perhaps in addition to a unificatory source of mathematical understanding, we ought to be looking to articulate a notion of non-causal determination.

In these comments so far I have had in mind primarily the mathematical explanation of mathematical phenomena, but I would also like improve my grip on apparently mathematical explanations of physical phenomena. For example, suppose that parents notice that when their child is behaving exceptionally badly, punishment is usually followed by some improvement in behaviour, whereas when their child is behaving exceptionally well and is rewarded, the subsequent behaviour is often somewhat worse. From this they infer that punishment is more effective than reward. This inference is probably unwarranted, because these are the patterns of behaviour one would observe if both punishment and reward were entirely inefficacious: the pattern can be explained simply by appeal to regression to the mean. Here we seem to have a lovely explanation of a physical phenomenon in terms of a mathematical fact.

Here is another example. You throw a bunch of sticks into the air with a lot of 'English', so that they tumble and spin as they fall. Now freeze the scene just before the lowest stick touches the ground. What you find is that appreciably more of the sticks are closer to the horizontal than to the vertical. Why is this? It is because of the broadly mathematical fact that there are more ways for a stick to be near the horizontal than there are for it be near the vertical. (Think of a single stick with a fixed midpoint: there are only two ways for it to be vertical, but indefinitely many ways for it to be horizontal. The asymmetry is preserved for near the vertical and near the horizontal.)

These cases of mathematical explanations of physical phenomena raise a number of interesting questions. Are they really non-causal explanations? Is the mathematical fact really doing the explaining? What if anything do these apparent cases of the mathematical explanation of physical phenomena tell us about the mathematical explanation of mathematical phenomena?

So far I have written (and until recently I have thought) that understanding is simply the flip side of explanation: 'understanding' is just the name for what you get from an explanation. But I have begun to think that this is too restrictive a notion of understanding, that while explanation is one route to understanding, it is not the only one. The more radical form of this thought is that there are forms of understanding that explanations do not provide. The less radical form is that the same forms of understanding that explanations provide may also be acquired by other routes. While not wishing to extend the notion of understanding so far as to deprive it of any interesting content, I am attracted to both forms. Thus working with a scientific theory may provide a kind of intellectual know-how that is tantamount to a form of understanding the phenomena, but one that is distinct from the kinds of understanding that explanations provide. As for the less radical thought, it looks like most of the characteristic cognitive benefits of explanation can be acquired by other means.

Take for example the necessitarian idea, the idea that an explanation may provide understanding by showing that the phenomenon had to occur. It seems that it is sometimes possible to show necessity and so to provide this kind of understanding without explaining. Thus we come to understand why gravitational acceleration is independent of mass by appreciating Galileo's wonderful thought experiment. Suppose that heavier things accelerate faster than light things, and consider a heavy and a light mass connected by a rope. Considered as two masses, the lighter one should slow down the heavier one, so the system should accelerate slower than the heavier mass alone. But considered as one mass, the system is of course heavier than the heavy mass alone and so should accelerate faster than the heavier mass alone. But the system can't accelerate both slower and faster, so acceleration must be independent of mass.

Because it is a *reductio*, the thought experiment does not itself seem to be an explanation. Nor does it appear that the route to understanding passes subsequently through an explanation that the thought experiment supports. Having absorbed Galileo's argument, I understand why acceleration must be independent of mass; but if you were to ask me to explain why acceleration is independent of mass, I cannot do it: the best that I can do would be to give you the thought experiment so that you can see for yourself. If, in the case of physical phenomena, we can acquire understanding without explanation, is the same true for mathematical understanding? This seems plausible, though the Galilean example may not make the case, since while in that physical instance coming to see the necessity may take one across the bridge from knowing to understanding, as I have suggested above this may not be so in purely mathematical instances, since here the necessity is already a part of mere knowledge.

Although I have already raised more issues about mathematical understanding than we will probably have time or inclination to pursue in the Symposium, I want to end by mentioning two more topics from my own work that might suggest points of comparison between physical and mathematical explanation. The first of these concerns the interest-relativity of explanation. It is a familiar and plausible claim that what counts as a good explanation is not determined by the phenomenon in question alone: it also depends on the interests and background knowledge of the person asking the question. For example, a good answer normally needs to give the inquirer something she doesn't already know.

In the context of a causal model of the explanation of physical phenomena, the need to take interest into account is clear in light of the density of causal histories. Behind every phenomenon there are innumerable causes, not all of which are explanatory. Thus while that computer crash does explain why my student's essay was late, the Big Bang doesn't, even though it is part of the causal history of every event. Moreover, the same cause may be explanatory for one person but not for another. Thus to take an example familiar from the literature, given that the only cause of paresis is untreated syphilis, but most people with untreated syphilis do not go on to contract paresis, one person may find it explanatory to be told that Smith contracted paresis because he had untreated syphilis while another person may rightly reject the explanation.

Some aspects of interest-relativity can be naturally analysed by giving more structure to the why-question. For many why-questions do not take the simple form 'Why P?' but rather have the contrastive form 'Why P rather than Q?' The choice of the foil Q makes a difference, and people with different interests may choose different foils. Thus if the real question is why did Smith rather than Jones contract paresis, where Jones did not have syphilis, then citing Smith's syphilis will be explanatory, but if the question is why Smith rather than Doe contracted paresis, where Doe also had syphilis, then it will not be. ('But Doe had syphilis too', your interlocutor will reply.) In my view, many of these contrastive questions generate a kind of triangulation that marks a distinction between explanatory and unexplanatory causes. Roughly speaking, what we need for explanation in these cases is a cause of P that 'makes a difference' between P and Q, and this can be seen as a cause of P for which there is no corresponding event in the case of Q. Smith's syphilis explains why he rather than Jones had paresis but not why he rather than Doe had paresis, because Jones did not have syphilis while Doe did. Does any of this carry over to mathematical explanations? Presumably here too there are various forms of interest-relativity. But is there a place for a contrastive analysis? And if so, how does the choice of contrast help to distinguish between explanatory and non-explanatory information?

Finally, I cannot forbear mentioning my interest in inference to the best explanation. This is the idea that scientists (and laypeople) often seem to use explanatory considerations as a guide to inference. They infer that a hypothesis is correct because although it is of course not the only hypothesis logically compatible with the evidence, if it were true then it would provide the best explanation of that evidence. If we want to articulate this thought, then one of the things we need to do is to say more about what 'best' is supposed to mean in the slogan 'inference to the best explanation'. For example, should we take 'best' explanation to mean the likeliest, that is the most probable explanation, or should we rather take it to mean the explanation that would if correct provide the greatest degree of understanding, that is the 'loveliest' explanation?

Likeliest might seem the obvious choice, since we want what we infer to have high probability. But I think this is the wrong choice in this context, since it would make the explanationist idea almost vacuous, reducing it to the claim that scientists infer what they take to be the most probable hypothesis. The initial attraction of the idea of inference to the best explanation was that it would illuminate scientists' inferential practices, but to say that scientists prefer what they take to be the most probable hypothesis is disappointing on this score. We get a much more interesting account of inference if we choose 'loveliest'. It is far from trivial to say that scientists tend to infer that the explanation that would if correct provide the greatest understanding is the explanation that is also likeliest to be correct: far from trivial, and maybe even approximately true. Those who find this line of thought attractive have their work cut out for them, since now they need to say what makes one potential explanation lovelier than another.

Does the idea of inference to the best explanation gain any purchase in the context of mathematical reasoning? This might seem unlikely, since inference to the best explanation is meant to provide a partial account of non-demonstrative reasoning, whereas at least the paradigm of mathematical reasoning is deductive. Who needs to appeal to the wishy-washy notion of inference to the best explanation when you have a proof? But things look different if we look at the context of discovery rather than of justification. For the heuristic of mathematical investigation is no demonstrative process, and it is possible that something like the idea of inference to the best explanation is applicable here. This is a question we might be able to answer, if only we can get a bit clearer about the nature of mathematical understanding.

Addendum on Peter Lipton's 'Mathematical understanding'

Stewart Shapiro

At the Castel Gandolfo Symposium, we had a most fruitful discussion of Peter Lipton's contribution, 'Mathematical understanding', in light of the other papers presented at the Symposium, and other work by him and the participants. His keen understanding and wit, together with his modest, disarming style, made the session especially interesting and compelling. With Lipton's sudden, shocking, and tragic death, there can be no update or revisions to his chapter. I hope here to provide some background to the work, and to locate it in some wider contexts.

Lipton was a philosopher of science, with special interests in the notions of explanation and understanding. His book, *Inference to the Best Explanation* (Lipton, 1991), is now close to being a classic—certainly required reading for anyone with even a passing interest in the topic. The second edition appeared in 2004. The focus of the book, and of much of his work, is on *scientific* explanation and understanding. In a preliminary, programmatic way, his Symposium contribution attempted to extend the concepts and ideas to mathematics.

As Lipton notes in his chapter, there is a clear and intuitive difference between *knowing that* a given proposition is true or, as he puts it, knowing that a given phenomenon occurs, and *understanding why* the proposition is true, or why the phenomenon occurs. Presumably, we already know 'that' before we ask 'why'; explanations are answers to 'why questions'. Aristotle (*Physics*, Chapter 3) wrote, 'Men do not think they know a thing unless they have grasped the 'why' of it.' This highlights the importance of explanation. Aristotle observes that we are often not satisfied merely knowing that something is true. We want an explanation of why it is true.

There is a long-standing literature on scientific explanation, going back at least 60 years (see, for example, Salmon (1990: 2006), covering the first 40 years or, for a more concise treatment, Woodward (2009)). The main competing models relate explanation to scientific laws (either universal or statistical), to causality, and to unification.

To summarize, on the first model, a scientific explanation of an event is a derivation of a description of the event from a scientific law (or laws), together with some initial conditions. In this sense, to explain an event is to show that, in some sense, the event was necessary—that it had to happen, given the laws of nature. On a second model, to explain an event is to say something relevant about its causal history. Lipton (1991: 2004) favors this model of explanation, with causality broadly construed, but he is open to other options. The third model ties scientific explanation to unification. On this picture, proper explanations serve to unify a number of apparently disparate phenomena, showing how they derive from a common source.

When it comes to explanation in mathematics, there are two sorts of issues. One is the phenomenon of explanation *within* mathematics. The intuitive distinction between knowing that a certain mathematical proposition is true and understanding why it is true seems to operate within mathematics as well as in science. Some proofs are thought to be explanatory and others not so. And Aristotle's observation holds there as well as in natural science and in ordinary life.

The other issue concerns cases where a mathematical fact is cited as an explanation of a non-mathematical event. Lipton gives two examples of this, both statistical. As he emphasizes elsewhere, there is always pressure to clarify exactly what the explanandum is or, as he puts it, what the contrast class is. Are we trying to explain why a given *pattern* tends to occur: why thrown sticks tend to be more horizontal than vertical; why kids tend to behave in certain ways after reward and punishment, etc. If so, then his point is that the explanation of the pattern often involves a mathematical theorem, such as regression to the mean, a measure-theoretic fact about spheres, or, in other cases, the central limit theorem, a staple of contemporary statistical theory. In each case, the theorem shows that the pattern in question is quite likely, given some background assumptions about probabilities.

The situation is a bit more complicated if the explanandum is a single *event*: why did more sticks end up horizontal this time? In principle, at least, one can explain this in terms of the initial velocities of the sticks, the air resistance at the time, etc., without invoking the geometrical fact that Lipton cites or anything about probabilities. But it is surely more pragmatic, and more illuminating, to note that the event at hand is an instance of a pattern that, according to the underlying assumptions about probabilities and some mathematics, was very likely to occur.

In any case, there are other, non-statistical examples where a mathematical fact is cited in an explanation of a non-mathematical event. Here is a rather simple-minded one: suppose that a child is given a number of rectangular tiles, all congruent to each other, and told to make a rectangular grid with them, using all of the tiles. After several attempts, she fails, and begins to wonder why she cannot complete her task. The explanation is that the number of tiles she was given is prime. For a more serious example, we all know that rain forms into drops. Why? The explanation, it seems, involves a detour through the notion of surface tension, together with the mathematical fact that a sphere is the largest volume that can be enclosed within a fixed surface area.

When it comes to mathematics, the first two models would seem to be out of place. We do speak of mathematical 'laws', such as the commutative law and the law of cosines, but that is probably only a manner of speaking. In science and ordinary discourse, it is common to distinguish a law (such as the law of gravity—all objects attract) from a merely accidental generalization (such as the fact that all US Presidents before the publication of this volume were male). That distinction does not seem operative in mathematics since, as Lipton notes, every truth in mathematics is necessary. More importantly, the use of 'laws' in mathematics does not seem to play a role in distinguishing explanatory proofs from non-explanatory ones. Presumably, everything in mathematics is derived from 'laws'. Nor does the notion of mathematical 'law' seem to play much of a role in the citation of mathematical facts in explanations of physical phenomena. Similarly, the notion of causality is also out of place in mathematics. There is no sense of a mathematical proposition being the 'cause' of a mathematical or non-mathematical proposition.

So among the standard models for scientific explanation, only unification seems to be applicable to mathematics. One of the main advocates of that model, Philip Kitcher (e.g., 1989) makes that point explicitly. Mark Steiner (1978, 1980) has written extensively on both sorts of mathematical explanation, and has developed an account that is in the vague neighbourhood of unification (but very different from Kitcher's). That was discussed extensively at the Symposium.

As noted, in his (2004) book, Lipton favours the causal model of explanation. At the Symposium, he tentatively suggested that causality may be just one of a family of dependence relations, and that the notion of explanation might be tied to dependence relations generally. This proposal has the potential to bring mathematics into the fold. The idea is that mathematical propositions stand in some sort of objective dependency relations to each other. If this is so, then one might maintain that explaining a mathematical proposition consists of showing what propositions it depends on. An explanatory proof would be one that reveals dependency relations among the premises and conclusion of the proof, while a non-explanatory proof would show that its conclusion is true without going via the dependency relations for the proposition.

We might go even further, and postulate that some physical events depend, in part, on mathematical propositions, such as the aforementioned examples concerning patterns in nature and theorems of statistics, tiling possibilities and theorems about prime numbers, and raindrops and theorems of geometry. If those are genuine dependency relations, then we can bring mathematical explanations of physical phenomena into the fold as well.

Gideon Rosen's contribution at the Symposium also invoked objective dependency relations between propositions, including, especially, propositions in different fields, including mathematics, although Rosen's interests are more in metaphysics than in the understanding of explanation. There was an extensive exchange between Rosen and Lipton, and the rest of us, on the nature of the various dependency relations, their objectivity, and their relevance for certain philosophical purposes.

Rosen reminded us that the eminent logician and philosopher of language, Gottlob Frege, also invoked objective dependency relations in mathematics (e.g., Frege, 1884: 1960). Frege often used epistemic terms for this, speaking of the 'justification', 'the proof', or sometimes 'the ground' of a proposition, but his rhetoric seems to be in the neighbourhood of present concerns. He took the dependency relation to be objective:

> ... we are concerned here not with the way in which [the laws of number] are discovered but with the kind of ground on which their proof rests; or in Leibniz's words, "the question here is not one of the history of our discoveries, which is different in different men, but of the connection and natural order of truths, which is always the same"
>
> (Frege (1884: 1960, §17); Leibniz, Nouveaux Essais, IV, §9)

Frege's dependency relation is also asymmetric: no proposition grounds itself, and if one proposition is a ground for another, the second is not a ground for the first. He tried to show that arithmetic and analysis are part of logic, by arguing that the basic propositions of those disciplines depend on general logical laws and definitions.

We can put Frege's project in the mould of Lipton's and Rosen's proposal, by suggesting that Frege was looking for the proper explanation of the propositions of arithmetic and analysis. He was not content with knowing *that* the propositions are true. Scepticism and fictionalism aside (see Mary Leng's chapter), there is no question that we do know these propositions. Frege was attempting to reveal *why* they are true. And that question is answered by revealing the propositions that arithmetic and analytic propositions depend on. His proposal is that the derivations of the basic principles of arithmetic and analysis that he provides are explanatory proofs.

As indicated by the title of his book (2004), one special focus of Lipton's thinking is on inference to the best explanation. The idea is that a scientist will often infer that a given proposition is true, or likely to be true, just because it would make for the best, or as Lipton puts it, the 'loveliest', explanation of some phenomenon. He closes his chapter with the suggestion that mathematics does not deploy inference to the best explanation:

> Does the idea of inference to the best explanation gain any purchase in the context of mathematical reasoning? This might seem unlikely, since inference to the best explanation is meant to provide a partial account of non-demonstrative reasoning, whereas at least the paradigm of mathematical reasoning is deductive. Who needs to appeal to the wishy-washy notion of inference to the best explanation when you have a proof? (p. 54)

He then briefly suggests that we shift from thinking about the context of justification, and focus on the context of discovery.

On the broadly Fregean picture sketched just above, however, we may not have to leave the realm of justification in order to find a role for inference to the best explanation in mathematics. The Lipton–Rosen–Frege idea is that some mathematical propositions rest on, or depend on, others. As Frege was aware, the regress cannot go on forever. Some propositions lie at the base of the foundational enterprise, and are not explained by anything. These are the axioms. Earlier in the essay, Lipton suggests that when it comes to proper axioms, there are no legitimate 'why-questions', or at least no legitimate answers to 'why-questions': 'Thus one might hold that mathematical understanding flows from the special epistemic status of mathematical axioms, and that these axioms are where the regress stops.'

But how are the axioms known? The traditional view is that the axioms are 'self-evident'. A full and complete understanding of an axiom immediately gives rise to justification for it. It seems to me, however, that this, traditional foundationalist view is not really tenable when it comes to modern mathematics. Some propositions presented as axioms are hardly obvious, and it is a bit of a stretch to say that a full and complete understanding of them will justify them (see Shapiro, 2009). Perhaps we can say, instead, that at least some axioms are chosen, not because of any intrinsic or self-evidence they may have, but because they make for a good, or, as Lipton might put it, lovely explanation of some of the theorems. It is a holistic picture. In a well-systematized branch of mathematics, the theorems are shown to depend on the axioms. In Aristotelian terms, the 'why' of the theorems lies ultimately in the axioms. When we turn to the axioms, and ask why they are true, or at least how it is that we know them, the answer is that they provide the best explanation of the theorems.

6
Creation and discovery in mathematics
Mary Leng

One important role for the philosophy of mathematics is to account for the phenomenology of the discipline, that is, to account for *what it feels like* to do mathematics. One aspect of this phenomenology is the sense mathematicians often have that they are discovering, rather than creating or inventing, the nature of mathematical reality. Given this aspect of mathematical practice, a natural assumption is that mathematicians are involved in the investigation of a mathematical reality that is independent of creative human decisions, and independent of our beliefs about that reality, in much the same way that physical scientists are involved in the investigation of a physical reality whose nature does not depend on us. If we accept this assumption, and the analogy on which it is based, then the question arises: 'What is the nature of this mathematical reality, and how is it possible for us to have knowledge of it?'

Taking seriously the analogy with physical science would suggest that mathematicians investigate a realm of mathematical *objects*, inquiring into the nature of numbers, sets, etc., just as physical scientists inquire into the nature of atoms. But if there is a mathematical realm of independently existing, non-physical objects, over and above the realm of physical objects we ourselves inhabit, then the question of how mathematical knowledge is possible becomes pressing. Our knowledge of the physical realm stems from our interaction, as physical beings, with that realm, but can mathematical knowledge be accounted for in an analogous way? Mathematician G. H. Hardy described mathematical discovery as observation of mathematical reality:

> *I believe that mathematical reality lies outside us, that our function is to discover or observe it, and that the theorems which we prove, and which we describe grandiloquently as our "creations," are simply the notes of our observations.*
>
> *(Hardy, 1940)*

But is this account of mathematical discovery as rooted in observation of a mathematical realm tenable?

Certainly, Plato thought that knowledge of the mathematical realm could be accounted for as resulting from a kind of direct 'observation'. According to Plato, mathematical objects belong to an eternal realm of Forms, which are directly perceived by the immortal, immaterial soul prior to its 'birth' as a flesh-and-blood human. In our physical incarnations, what mathematical knowledge we have is had by recollection of this direct experience of the Forms (Plato, *Meno*, 81d–86c). Our theorems then report what our mathematical inquiry has enabled us to remember about our earlier direct observation of the mathematical realm. However, many would find this account of our knowledge of mathematical reality hard to swallow, requiring as it does some powerful and potentially problematic assumptions about mind and body. Perhaps something like Plato's picture could be defended on the basis of an inference to the best explanation: implausible as it may sound, it must ultimately be accepted as the only good way of accounting for the phenomenology of mathematical discovery. However, before opting for this solution, it is worth examining the phenomena in question, to consider what alternative explanations might be available.

Just what is it that mathematicians seem to discover? The discoveries Hardy mentioned were of theorems, presumably within established mathematical theories. But mathematicians also, of course, create/discover new mathematical theories, within which theorems can be proved. We must, then, consider whether either of these kinds of discoveries are best viewed as discoveries of the nature of a realm of mind-independent, non-spatiotemporal mathematical objects.

One might think that the discovery of entirely new theories provides the best evidence for such a mathematical realm. After all, once the axioms or basic presuppositions of a theory are in place, our discovery of mathematical theorems is discovery of *what follows from* those presuppositions, and at least at first glance, this kind of 'what if' inquiry, into what would have to be true if our mathematical axioms were true, does not require that our mathematical axioms *are* in fact true of some underlying mathematical reality (though more on this later). On the other hand, in developing new mathematical theories, mathematicians often have a sense of discovering the basic assumptions of these theories, as assumptions that truly describe an important corner of mathematical reality, rather than simply plucking theoretical hypotheses out of the air with a view to inquiring blithely into their consequences. Surely this aspect of the phenomenology of mathematical practice provides the strongest evidence for an independently existing realm of mathematical objects?

In fact, I will argue, the phenomenology of theory development is actually easier to account for from an anti-Platonist perspective than is the phenomenology of mathematical proof within theories. If accounting for the phenomenology of mathematical discovery requires us to posit any kind of 'reality' to ground our mathematical judgements, this reality is not a realm of mathematical objects, but rather, I claim, a realm of objective facts about logical consequence. Insofar as we are concerned with understanding the sense of discovery that is present in mathematical theorizing, the real puzzle to be accounted for (a puzzle that, in fact, already arises even when one considers ordinary empirical reasoning) concerns what Wittgenstein called 'the hardness of the logical must' (Wittgenstein, 1953: 2001, I, §437), and not the existence of a realm of mathematical objects.[1]

Let us start, though, with theory development. It is certainly true that the choice of assumptions for a new mathematical theory is usually far from arbitrary, and indeed that the development of appropriate theoretical assumptions is often rightly viewed as a significant and extremely nontrivial achievement. But does this require us to view the development of new mathematical theories as the description of an independently existing realm of mathematical objects?

The evidence of mathematical practice, I think, speaks against this, suggesting as it does constraints on our theory development that account for our sense of discovery without requiring us to posit a realm of mathematical objects to be discovered. For, very often, mathematical theories are developed as solutions to problems we have set ourselves, where the constraints of the problems are enough to narrow down the range of options that could count as an appropriate solution (often even pinning down a unique solution). Take, for example, W. R. Hamilton's discovery of the quaternions which, he tells us, 'started into life, or light, full grown, on the 16th of October, 1843' (quoted in Tait, 1866, p. 57). Was Hamilton's moment of inspiration, the discovery of the equation $i^2 = j^2 = k^2 = -1$, which he excitedly carved into the stone of Brougham Bridge, a sudden recollection of a truth contemplated by his pre-embodied soul?

In fact, as Hamilton's own description of his 15-year struggle to develop rules of addition and multiplication for a three-dimensional analogue of two-dimensional complex numbers shows, his moment of inspiration is better viewed as a sudden realization of what had to hold, given the constraints he had set himself. Hamilton's aim was to discover laws for multiplying triplets of the form $x + iy + jz$, along the lines of the laws for multiplying pairs $x + iy$, where j was to be viewed as a square root of -1 distinct from i. A constraint he set himself was to satisfy the 'law of the moduli': the modulus of the product of two triplets should equal the product of the moduli of the two triplets taken separately. That is, if $(a + ib + jc) \cdot (x + iy + jz) = u + iv + jw$, then the law of the moduli would require that $(a^2 + b^2 + c^2) \cdot (x^2 + y^2 + z^2) = u^2 + v^2 + w^2$. This constraint, it turns out, is impossible to satisfy if multiplication is assumed

---

[1] This is not to say that there are no aspects of mathematical practice that might require us to assume the existence of mathematical objects. Indeed, if we turn to the question of the applicability of mathematics, considerations of scientific confirmation might require us to hold, not only that there are objective facts about the consequences of our hypotheses concerning mathematical objects, but also that some of those hypotheses are in fact true.

to have its usual properties of commutativity and associativity. Dropping the commutativity constraint, however, showed Hamilton that there was room for a solution in some special cases, so long as the product ij was equal to $-ji$, and that neither were equal to 0. This effectively forced Hamilton's hand, leading him to discover that the only possible value for the product ij was a third imaginary, k.

The exact problem Hamilton originally set himself had no solution. However, the system of quaternions emerged as the best way to fit as many of Hamilton's original constraints as possible. In fact, if we set the problem in advance as that of extending multiplication to a system of numbers based on n square roots of -1, preserving associativity and the law of the moduli, it is a matter of logical consequence, rather than a mathematical 'matter of fact', that only three solutions are possible, these being $n = 0$ (the real numbers); $n = 1$ (the complex numbers), and $n = 3$ (the quaternions).

In more pedestrian cases of mathematical theory development than Hamilton's, we often find that axiomatizations are governed by the constraint to come up with a system of assumptions that will pin down essential features of an already familiar mathematical or empirical system. For example, Euclid's axioms for geometry were intended to pin down truths about points and straight lines in physical space, and were 'discovered' through examination of the question of what had to be assumed in order to prove many other results believed to be true of points, lines and geometric shapes. And aside from physical interpretations, the development of axioms for mathematically familiar objects is also commonplace—as, for example, in the development of the Dedekind-Peano axioms, where the axiomatization is constrained by the requirement that the structure axiomatized be an ω-sequence. Indeed, axiomatizations of mathematically familiar objects often appear initially as theorems—for example, the axioms of my own favourite corner of 'mathematical reality', C^*-Algebras, first appeared as a part of the Gelfand-Naimark *theorem*, which showed that those axioms pinned down up to isomorphism subalgebras of the algebra $B(\mathcal{H})$ of bounded operators on a Hilbert space, structures which were already of independent mathematical interest.

We are constrained, in all of these cases, by what is known already about the system to be captured, and this constraint of course leads to the sense that, in coming up with a formal theory, we are 'getting something right'. In such cases, our axioms might strike us as being 'true', rather than merely convenient or interesting, but this sense of correctness is explicable even if we do not invoke the existence of a realm of objects for our axioms to be true *of*. What counts as getting something right in these cases is dependent on the constraints we have set ourselves (which require us to pin down a structure satisfying various assumptions). We need not (though we may) think of these constraints as being imposed by an independent realm of objects about which our theories must assert truths. Rather, as consequences of a collection of initial assumptions or constraints, new axiomatizations can come about in much the way that theorems do within the context of already established theories. In each case, the area of mathematical discovery that really matters seems to be the discovery of consequences of one's assumptions.

Let us turn, then, to the kind of mathematical theorizing that takes place against a backdrop of accepted assumptions (such as axioms). In the paradigm cases, such reasoning is deductive, and amounts to the proving of theorems from axioms, although there is also some room for the use of abductive reasoning in such contexts: mathematicians may reason that, given their assumptions, such and such a result is *likely* to be true. Sticking for now with the central case of deductive proof from axioms, we can consider what conclusions can be drawn from the sense mathematicians often have that in such activity they are involved in *discovery* rather than *creation* of mathematical results. Are mathematicians who are engaged in proving theorems really discovering the already determined consequences of their assumptions, or could it be the case that, despite the strong sense of discovery, they are actually involved in the creation of links between axioms and theorems that were not already, in some sense, 'out there'? If mathematicians *are* engaged in discovery rather than creation, what implications does this have for our view of the nature of mathematics. In particular, is this discovery discovery of a realm of mind-independent mathematical objects? On the other hand, if we do choose to view deductive proof from axioms as a matter of creation, rather than discovery, can this be reconciled with the felt objectivity of mathematical proof, and, indeed, the applicability of mathematical reasoning?

A natural line of thought takes it that, yes, deductive mathematical reasoning is objective, leading us to the discovery of logical consequences of our mathematical assumptions. But such objectivity has nothing to do with an independent realm of mathematical *objects*, but is entirely a result of the objectivity of logic. After all, in reasoning to a theorem P on the basis of mathematical assumptions $A_1, \ldots A_n$, we prove not that P is true, but rather, that *if $A_1 \& \ldots \& A_n$, then P*. This conditional claim does not assert the existence of any mathematical objects; its truth, we may suppose, rests solely on the fact that P is a logical consequence of $A_1, \ldots, A_n$. According to this way of thinking, there is nothing particularly problematic about the felt objectivity of mathematical proof from axioms—it is simply a special case of the objectivity of *any* deductive reasoning, and depends solely on the objectivity of the 'following from' relation. Furthermore, as we have seen, a similar story can be told about the kind of reasoning that leads to the development of new mathematical theories: although this reasoning does not start from axioms, it is nevertheless governed by *logical* constraints established by the consequences of our preformal mathematical assumptions and/or desiderata.

There is, however, a problem with this comfortable-seeming position, which arises once we consider what we mean by the claim that P follows logically from $A_1, \ldots, A_n$. We surely do not mean by this that P can be *derived* from $A_1, \ldots, A_n$ using an accepted collection of rules of inference. For one thing, we know of cases where this analysis falls short of capturing our usual notion of logical consequence: take the second-order Peano axioms for arithmetic. Gödel's first incompleteness theorem tells us that, for any (standard) collection of inference rules we can come up with, there will be a sentence G in the language of (2nd order) Peano arithmetic which follows logically from those axioms but which is not derivable using our chosen collection of rules. But even without the worries that Gödel's theorem brings, one should be wary of an analysis of logical consequence which rests facts about logical consequence on a chosen collection of inference rules; after all, what makes a collection of inference rules a good one is, presumably, that those rules respect facts about logical consequence, and not vice versa.

All of which is simply to say that the relevant notion of logical consequence underlying the objectivity of mathematical reasoning is that of *semantic*, rather than *syntactic* consequence. In the relevant, semantic, sense, P is a logical consequence of $A_1, \ldots, A_n$ if and only if it is not logically possible for P to be false while $A_1, \ldots, A_n$ are all true. But this analysis simply replaces one undefined logical notion (logical consequence) with another (logical possibility). If our question is what grounds the objectivity of logical consequence, then surely an analogous question arises for logical possibility: can we say anything more about what the logical possibility or impossibility of a sentence amounts to?

Here is where the difficulty for our comfortable view of the objectivity of mathematical reasoning arises. For, arguably, the best analysis of logical possibility available is *mathematical*: a sentence P is logically possible if there is a set theoretic model in which that sentence is interpreted as a truth. On this analysis, P is a logical consequence of $A_1, \ldots, A_n$ if and only if, in all models which make $A_1, \ldots, A_n$ true, P is also true. If this analysis is correct, then the existence of objective facts concerning logical consequence comes down to the existence of a realm of mathematical objects (set theoretic models). So the objectivity of mathematical discovery is, after all, dependent on an objective realm of mathematical objects, and we are led back to the difficulty of explaining our knowledge of such things.[2]

Is there an alternative analysis of logical possibility available? One might try to eschew abstract mathematical objects in favour of logically possible concrete worlds. But even if we could make sense of the notion of a logically possible world in such a way as to respect our intuitions concerning logical possibility, if it is facts about these worlds that ground the objectivity of mathematical inference, we will still have difficulty explaining how we could have knowledge of the following-from relation, since such worlds are presumably spatiotemporally isolated from our own. Since we do seem to know some facts about what follows from what, we should be wary of any analysis of the

---

[2] Georg Kreisel reportedly maintained that the question of realism in mathematics amounts to 'the question of the objectivity of mathematics and not the question of the existence of mathematical objects.' (Putnam, 1975: 1979, p. 70). If this analysis of mathematical objectivity is correct, then these two questions cannot, after all, be separated.

following-from relation that grounds such facts in matters which seem forever beyond our grasp.

Such concerns might lead us to abandon all attempts to *reduce* logical possibility to something more basic. Indeed, drawing from a discussion of Kreisel's (1967), Hartry Field (1984: 1989, 1991), has argued that we should view logical possibility as a distinct notion from the related formal notions of deductive and model theoretic consistency. We can learn about logical possibility via derivations and models: we know that, if (in an accepted derivation system) we can derive a contradiction from a sentence *S*, then *S* is not logically possible, and that, if (in an accepted set theory) we can find a model in which the sentence *S* is interpreted as a truth, then *S* is logically possible. But (on the Kreisel/Field view), logical possibility is a distinct notion from the related deductive and model theoretic notions, and should not be thought of as reducible to either. Rather, Field suggests, we should see 'it is logically possible that' as a unary logical operator that is no more in need of 'reduction' than is the unary operator 'it is not the case that'. Both, Field thinks, should be explicated through specification of their inferential role, rather than via a reduction to something more primitive. Note, then, that this account accepts the existence of irreducible modal facts grounding the objectivity of mathematical reasoning. While avoiding commitment to the existence of abstract mathematical objects, this account still requires us to accept a kind of reality underpinning our mathematical discoveries (albeit a 'realm' of modal facts, rather than of abstract mathematical objects). And once more we will need to ask what it is about us as humans that allows us to have knowledge of these modal facts.

But perhaps there is another response to the phenomenology of mathematical discovery: perhaps we can accept the *feeling* that our judgements concerning logical consequence have an objective ground that is independent of human decisions, but hold that this sense of objectivity is nevertheless an illusion. This is the approach that Wittgenstein takes in his conventionalist approach to mathematics. According to Wittgenstein, despite appearances, 'The mathematician is an inventor, not a discoverer' (Wittgenstein, 1956: 1978 I, p. 167). In *proving* mathematical theorems, we do not *discover* the consequences of our mathematical hypotheses, but rather, *decide* to accept the conclusion proved as a new consequence in our theory. Far from teasing out the content of our mathematical concepts,

> the proof changes the grammar of our language, changes our concepts.
> It makes new connections, and it creates the concept of those connections.
> (It does not establish that they are there; they do not exist until it makes them.)
>
> (Wittgenstein, 1956: 1978, III, p. 31)

Perhaps, then, despite appearances, there are no objective facts about logical consequence in mathematics, just the results of human decisions that could always have gone differently?

Perhaps this view could be sustainable if individual mathematical theories were entirely isolated from one another, so that 'decisions' made within one theory would not clash with decisions made elsewhere. Indeed, Wittgenstein himself thinks that cross-theoretical links are themselves a matter of decision, that, for example, it is a matter of choice to embed the natural numbers in the integers and so on (see, e.g., Waismann, 1979, pp. 34–6). But here the phenomenology of mathematical discovery speaks strongly against the conventionalist position, as Friedrich Waismann notes in explaining (in his (1982) paper, 'Discovering, Creating, Inventing') his own abandonment of Wittgensteinian conventionalism. There are just too many examples of theorems proved in one mathematical context bearing out (and even illuminating) the conclusions drawn in other areas. Waismann's example is of a result concerning the real numbers that receives an explanation once the reals are embedded in the complex numbers. The Taylor series expansion

$$\frac{1}{1+x^2} = 1 - x^2 + x^4 - x^6 + \cdots$$

converges for $|x| < 1$, but diverges for all other real values of x. Once we embed the reals in the complex numbers, the behaviour of the function $\frac{1}{1+z^2}$ in its real portion is explained by the fact that the complex function has singularities at $z = \pm i$, together with a theorem of complex analysis which tells us that any power series expansion only converges within a circle of radius R about the origin, and diverges elsewhere. Given these facts about *complex* functions, the real-valued function could not have behaved other than it did. Far from being a matter of human convention, such results seem determined independently of our choices. It is, as Waismann remarked, as if the real function *already knew* that the complex numbers were there.

Related to this issue (which we might call the applicability of mathematics within mathematics), is the phenomenon of the applicability of mathematics to nonmathematical questions. We are abundantly aware of cases where mathematical reasoning is used to derive empirical predictions, and where these predictions turn out to be correct. One view on the applicability of mathematics takes the applicability of mathematics to reside in structural similarities: a mathematical theory is (sometimes) applicable to a nonmathematical phenomenon because the nonmathematical reality is structurally similar to some portion of the structure described by the mathematical theory in question. But if in mathematical reasoning we were simply freely *deciding*, at each step, what we will take to be true of the objects of a given mathematical theory, it is surely entirely mysterious how these free decisions so regularly result in accurate predictions.

Both of these phenomena, then, speak against the radical anti-objectivist account of mathematical reasoning. What, then, does the phenomenon of mathematical discovery have to tell us about mathematical reality? Not, I think, that our mathematical theorems are answerable to an independent realm of mathematical *objects*. But if we do not accept Wittgenstein's extreme conventionalism, at the very least we must accept that our mathematical discoveries are underpinned by objective facts about logical consequence. And if we wish to hold, with Kreisel, that the problem that concerns us is ultimately 'not the existence of mathematical objects, but the objectivity of mathematical statements' (Dummett, 1978, p. xxviii), then we will have to accept that the relevant facts concerning logical consequence do not reduce to facts about mathematical models. If we want to understand mathematical discovery, then, we must consider from where the objectivity of these facts might arise.

Comment on Mary Leng's 'Creation and discovery in mathematics'

Sensing objectivity

Michael Detlefsen

———————

It is not uncommon for those experienced in doing mathematics to see it as an activity of discovery or observation rather than innovation or creation. G. H. Hardy and Kurt Gödel were among prominent twentieth-century mathematicians who believed so.

Leng agrees with Hardy and Gödel that a sense of discovery is a significant part of our mathematical experience. She believes in addition, though, that the most convincing 'felt objectivity' is one concerning facts of logical consequence.

Things might rest there but for the fact that, in Dr. Leng's view, the most compelling treatment of logical consequence is that given by contemporary model theory, and judgements concerning models involve us as much in difficulties concerning knowledge of abstract objects as views of 'felt objectivity' not restricted to logical consequence. She thus concludes that the 'felt objectivity' of mathematics cannot unproblematically be attributed to the objectivity of logical consequence. This notwithstanding, she does not see a clearly preferable alternative.

This may under-represent the difficulties of accepting the 'felt objectivity' of logical consequence as a datum for mathematical epistemology. In the first place, it does not address the possibility of different conceptions of logical consequence. If we include practising intuitionists and other kinds of constructivists among those having relevant mathematical experience, there will not be agreement on what the compelling instances of logical consequence are. Indeed for some (e.g., Brouwer), there will not be agreement on the importance of *any* judgements of logical consequence, intuitionist or classical, to what is rightly counted as *mathematical* thinking.

Greater attention might also have been given to what it is that is supposedly felt to be objective in judgements of consequence. Leng says it is semantical fact, but there are other possibilities too. As a mathematical activity, proving is intended to bring about a certain response in a certain audience. That this is so suggests that judgements of consequence ought ultimately to reflect what the prover judges to be the inferential standards of the intended audience. Whether those standards are best specified in semantical terms or in terms of comportment with trusted non-semantically specified rules is a matter on which there is room for deep disagreement.

Finally, the complexities of 'felt objectivity' may be underestimated. Gödel wrote of propositions 'forcing' themselves on us as being true. Is this what 'felt objectivity' is supposed to come to? Or is 'forcedness' only a part of 'felt objectivity'? If the former, then felt objectivity does not seem to provide much evidence for mind-independence. Innate dispositions and even conditioning brought about by training can surely give rise to feelings of being 'forced'. In his *Grundgesetze* (Frege, 1903: 1962, II, §142), Frege rightly warned against taking such feelings of compulsion as indications of truth. As he noted, one 'need only use a word or symbol often enough, and the impression will be produced that this proper name stands for something; and this impression will grow so strong in the course of time that in the end hardly anybody will have any doubt about the matter.'

If felt objectivity is only forcedness, then, it is not a potent indicator of truth or objectivity. If, on the other hand, it is more, then we need to be told what its other elements are and how they give surer evidence of mind-independence than mere forcedness does.

7
Discovery, invention and realism: Gödel and others on the reality of concepts

Michael Detlefsen

Introduction

This chapter is an investigation into the question whether there are features of our acquisition of mathematical knowledge that support a realist attitude towards mathematics. More particularly, it is a reflection on the reasoning which moves from the claim that

    I. mathematicians are commonly convinced that their reasoning is part of a process of discovery, and not mere invention,

to the claim that

    II. mathematical entities exist in a noetic realm to which the human mind has access.

For convenience, I'll refer to this as the *original argument*.

The use of the term 'noetic' in II calls for brief comment. Traditionally it has been used to signify a type of apprehension, *noēsis*, which is characterized by its distinctly 'intellectual' nature. This has generally been contrasted to forms of *aisthēsis*, which is broadly sensuous or 'experiential' cognition, or *intuition*. There is interest, and difficulty, in determining more exactly the ways in which an intellectual experience of a supposedly non-sensuous reality might resemble and might also contrast with sensuous experience of material objects. This is where a great deal of the difficulty concerning the content of 'noetic' will be met. The terms of such a comparison will therefore be one of our chief concerns.

Experience and involuntariness: background

For Plato, the objects of *noēsis* were the Forms, which he believed to be manifested by experience while also transcending it. The empiricists, on the other hand, generally associated intellectual apprehension with apprehension of concepts or *ideas*. These, when legitimate, were mental representations obtained by abstraction from sensory experience.

Kant famously emphasized a distinction between two types of representations, intuitions (*Anschauungen*) and concepts (*Begriffe*). Among their important differences, he maintained, was one concerning the extent to which they are within the power of the judging agent to control. Concepts were taken to be spontaneous (see Kant, 1781: 1990, A50–51/B74–75), or capable of being brought into existence by a judging agent's own intellectual initiative (*selbst ausgedachten*, op. cit., A639/B667). Intuitions were not. In the end, however, Kant required that genuine or legitimate concepts be consistent (*nicht selbst widersprechen*, op. cit., A150/B189). The freedom to create or generate them was thus a constrained freedom.

This notwithstanding, Kants saw a great difference between our control over concepts and our control over intuitions (see op. cit., A19/B33, B132). He regarded intuitions and their relations as *given* (*gegeben*) (see op. cit., A19/B33, B132) and not under the spontaneous productive control of our minds. Concepts, on the other hand, could be thus produced. There was therefore no guarantee that they be exhibited by any object(s).

> ... *even if our judgment contains no contradiction, it may connect concepts in a manner not borne out by any object, or in a manner for which no ground is given ... and so may still, in spite of being free from all inner contradictions, be either false or groundless.*
>
> *Kant (1781: 1990, A150; B190)*[1]

Kant's acceptance of this asymmetry between concepts and intuitions provides an interesting point of contrast to more recent views of the nature of concepts. Specifically, it seems to be sharply at odds with the view of concepts presented by Gödel in various foundational writings of the 40s, 50s and 60s (see Gödel, 1947: 1990). It is to Gödel's view(s) that I now turn.

Gödel agreed with Kant in regarding the involuntariness of a representation as a mark of its objectivity. However, whereas Kant regarded our use of mathematical *concepts* as essentially creative or voluntary,[2] Gödel regarded it as distinctly *in*voluntary. Contrary to Kant, he thus maintained that

> ... *despite their remoteness from sense experience we do have something like a perception also of the objects of set theory, as is seen from the fact that the*

---

[1] See Kant (1781: 1990, Bxxvi) for similar remarks.

[2] As Kant put it:

> ... *I can think* (denken) *whatever I want, provided only that I do not contradict myself, that is, provided my concept (Begriff) is a possible thought*

axioms force themselves upon us as being true. I don't see any reason why we should have less confidence in this kind of perception, i.e., in mathematical intuition, than in sense perception.

Gödel (1947: 1990, p. 268)

Indeed Gödel claimed more than the involuntariness of our apprehension of concepts. He believed that it has a perception-like, but still non-sensory character. He maintained as well that the cognitions it yields are in important ways independent of both the *voluntary acts* and *involuntary dispositions* of our minds.

I am under the impression that . . . the Platonistic view is the only one tenable. Thereby I mean the view that mathematics describes a non-sensual reality, which exists independently both of the acts and the dispositions of the human mind and is only perceived, and probably perceived very incompletely, by the human mind.

Gödel (1951: 1995, pp. 322–23)

Despite these realist convictions, however, Gödel conceded certain points to the conventionalist. Specifically, he believed, they were right to think that mathematics is about concepts rather than physical or psychical items (see Gödel, 1951: 1995, p. 320). They were right too, he said, to believe that mathematical truths in some sense owe their truth to the meanings of terms—specifically, to the concepts expressed by terms (loc. cit.). Where the conventionalist went wrong, he believed, was in taking these meanings to be determined by conventions (ibid.). The truth as he saw it was rather that

these concepts form an objective reality of their own, which we cannot create or change, but only perceive and describe.

Gödel (1951: 1995, p. 320)

In Gödel's view, then, mathematical concepts are discovered and not created by acts of convention or other mental acts or dispositions. Similarly for truths concerning them (loc. cit.). The chief evidence of this, in his view, was the involuntariness with which truths concerning mathematical concepts 'force' themselves on us as being true. This, to Gödel, was signal indication of their independence from our creative capacity, a capacity he took to mainly

---

(möglicher Gedanke). *This suffices for the possibility of the concept* (Begriff), *even though I may not be able to answer for there being, in the sum of all possibilities, an object* (Objekt) *corresponding to it. Indeed, something more is required before I can ascribe to such a concept objective validity* (objektive Gültigkeit), *that is, real possibility* (reale Möglichkeit); *the former possibility is merely logical. This something more need not be sought in the theoretical sources of knowledge* (theoretischen Erkenntnisquellen); *it may lie in those that are practical.*

*Kant (1781: 1990, Bxxvi, note *)*

be constituted by our mental dispositions and our various abilities to perform voluntary mental acts.

Gödel's phenomenological argument

Gödel's variant of the original argument thus emphasized what he took to be a broad phenomenological feature of our mathematical experience—namely, the involuntariness of our basic mathematical knowledge. Unlike the version of the original argument set out at the beginning, it attached little if any significance to the mere fact (if it is a fact) that mathematicians are *commonly convinced* that they discover rather than invent.

The core element of his argument was thus the claim that

1. Propositional contents concerning mathematical concepts are imposed or 'forced' on us as being true in a manner similar to that in which propositional contents are impressed on us by sensory experience.

He seems further to have believed that

2. This imposed character is best explained by seeing it as a consequence of a perception-like experience of a realm of beings whose existence and characteristics are independent of our mental acts (e.g., acts of convention or stipulation) and dispositions, both individual and generic.[3,4]

From this, he suggested, we may rightly infer that

3. Mathematical concepts exert a non-sensory cognitive influence on us, and their existence and properties are independent of our mental acts and dispositions.

The most plausible account of our overall mathematical experience thus implies that

4. Mathematical beliefs are about objectively existing things that we discover rather than invent or create (e.g., by acts of stipulation or convention).

This, in sum, is Gödel's argument, an argument which provides one type of more extended articulation of the original argument. What I find most interesting and distinctive about it is its appeal to a supposed phenomenon

---

[3] Gödel expressly denied, however, that our perception-like experience of mathematical truth was or was at bottom based on sensory perception. It was perception-like only in that it was given to or forced upon us in a manner akin to that in which the contents of sensory perception are given to or forced upon us.

[4] In saying that mathematical concepts (at least some of them) are 'independent' of our mental acts and dispositions, Gödel seems to have meant that (i) mathematical concepts exist and would continue to exist even if our mental acts and dispositions did not, and (ii) no change in our mental acts or dispositions would automatically result in a change in the properties of those concepts.

of 'forcedness' concerning our mathematical judgements (or at least some of them), and the idea that this phenomenon serves as evidence for the external reality of mathematical concepts. In the next section I'll consider Gödel's reasoning more carefully and also attempt to clarify certain ways in which it differs from earlier arguments for the reality of mathematical entities.

'Forcedness' as an indicator of reality

Gödel's language, especially his statement that mathematical propositions 'force themselves upon us as being true' (Gödel, 1947: 1990, p. 268) suggests a broadly phenomenological type of reasoning. More specifically, it is reasoning that takes its lead from what it supposes to be our experience of the evidentness of certain propositions.

Gödel seems to have been particularly concerned with the evidentness of basic or primitive mathematical propositions—propositions the forcedness of which does not seem to be explicable by their being perceivedly implied by (other) forced propositions. This at any rate is what he seems to have had in mind when he noted not only that certain *propositions* of set theory force themselves on us as being true, but that its *axioms* do.[5]

Gödel seems to have thought that proper sensitivity to the particulars of our experience of evidentness in mathematics will reveal that at least some propositions are 'forced on' us in a manner similar to that in which sensory propositions are 'forced on' us by sensory experience. He seems to have seen this, moreover, as indicative of an external mathematical reality to which we have broadly 'experiential' access.

> There exists, unless I am mistaken, an entire world consisting of the totality of mathematical truths, which is accessible to us only through our intelligence, just as there exists the world of physical realities; each one is independent of us, both of them divinely created.
>
> Gödel (1951: 1995, p. 323)

---

[5] Gödel's exact statement was 'the axioms [of set theory] force themselves upon us as being true.' 'The' axioms? Was Gödel assuming that there is a unique, or perhaps uniquely best, axiomatization of set theory? Not necessarily. Given a set-theoretic language L, there is no necessary incoherence in believing that (i) there is an identifiable set Π of propositions formulable in L whose evidentness does not derive from that of other propositions formulable in L, and also that (ii) not all elements of Π need or even ought to be taken as axioms of an axiomatization of set theory. There might, for example, be logical overlap between elements of Π that would make it unnecessary or even undesirable to take all of them as *axioms* of an axiomatization of set theory. There might also be alternative ways of thinking about sets, or different concepts of set, that would divide the elements of Π in such ways as would associate certain elements of Π with certain concepts or ways of thinking and not others.

Such reasoning, of course, raises many questions. Among these are:

Q1: How reliable and revealing an indicator of external reality is our experience of the forcedness of sensory propositions?

Q2: How reliable and extensive is the asserted analogy between the forcedness of basic mathematical propositions and the forcedness of basic sensory propositions as indicators of external realities?[6]

Q1 queries the evidential connection between the property of sensory judgements called forcedness and the existence of an external source responsible for it. Affirmation of such a connection would seem to require belief that forcedness of sensory judgement is best explained by some sort of transfer of energy—specifically, transfer *to* a sensory agent *from* a sense-stimulative source external to her.

Understood this way, is phenomenal forcedness of sensory judgements a reliable indicator of external reality? Here, I think, we need to distinguish two types of such reliability. One is what I'll call *existential* reliability, or reliability concerning the existence of an external source for an experience of forcedness. The other is what I'll call *attributive* reliability, or reliability concerning the characteristics of a supposed external source forced on us in a sensory judgement.

Exceptions to both existential and attributive reliability are, of course, familiar from the literature on sense perception and I'll not go into them in any detail here. Rather, I'll simply note that quasi-sensory experiences such as certain hallucinations provide cases where an experience of forcedness is not an existentially reliable indicator of an external reality. Similarly, well-known cases of optical illusion raise similar concerns regarding attributive reliability.

In addition to these concerns, there are three others I'll mention. The first has to do with our understanding of 'forcedness'. What did Gödel mean when he said that our mathematical experience includes the experience of propositions' forcing themselves upon us as being true? A natural interpretation would include the following implication: *P*'s being forced upon us as true implies that we form a belief that *P*.

This raises an important question concerning Gödel's supposed mathematical 'perception'. The reason why is that sense perception does *not* seem to sponsor the above implication. That is, having a sensory experience as of *P* (e.g., an experience as of one line segment's being longer than another) does not seem to imply that we form a belief that *P*. There are well-known illusions (e.g., the Ponzo and Müller-Lyer illusions) in which we experience one line segment's being longer than another but do not *believe* it to be so.

---

[6] More basic than either Q1 or Q2, of course, is the difficult question of how to make sense of the notion of a sensory proposition. I won't address this problem here, though, since it is Q1 and Q2 and their focus on the phenomenon of forcedness that are my chief concerns.

In sensory perception, then, we can have an experience as of P but not judge or believe that P. Sensory perception seems to be a form of apprehension in which not all contents it presents are presented as true. Is the same true of Gödel's mathematical 'perception'? That is, can mathematical perception in some sense 'give' appearances which, despite there being given in this way, are nonetheless not forced on us as being true?

I don't know the answer to this question, but it's not the answer that is my chief concern here. Rather, it is the question itself. It shows, if I am not mistaken, a certain type of difficulty posed by Q2, a difficulty that will confront any attempt such as Gödel's to exploit an analogy between sense perception and a perception-like form of apprehension in mathematics.

Nor is this the only such difficulty or the most serious. More troublesome, I think, are problems raised by cases of sensory illusion where false contents are forced on us as being true. A well-known illustration is Adelbert Ames' 'Distorted Room'. Viewed in such a way as to eliminate stereoptic information (e.g., viewed through a peephole), the room looks like an ordinary 'cubical' room with rectangular windows, a flat, rectangular, level floor and rectangular walls of equal and uniform height and depth.

When commonplace objects (e.g., an adult human being of normal size and a smaller child of normal size) are placed on opposite sides of the room and photographed, and the photographs are merged into a single image, strange appearances occur. The child, for example, will appear to be much larger than the adult.

The truth, of course, is that Ames' room is not, despite its appearance, an ordinary cubical room. Its seemingly rectangular windows, walls and floor are trapezoidal, its walls are not of uniform height and depth and the floor is not level. That it looks normal, and continues to look normal when familiar types of objects are placed in it suggests the influence that *dispositions* can have on perception.[7] As R. L. Gregory described it:

> *Evidently we are so used to rectangular rooms that we accept it as axiomatic that it is the objects [human bodies] which are odd sizes rather than that the room is an odd shape. But this is essentially a betting situation—it could be either, or both, which are peculiar. Here the brain makes the wrong bet, for the experimenter has rigged the odds. Indeed . . . the most interesting feature of the Ames Distorted Room is its implication that perception is a matter of making the best bet on the available evidence . . . wives do not see their husbands as distorted by the Room—they see their husbands as normal, and the room its true queer shape . . . [Also] Familiarity with the room, especially through touching its walls . . . does gradually reduce its distorting effect on other objects, and finally it comes to look more or less as it is—distorted in fact.*
>
> Gregory (1969, pp. 180–181, brackets added)

---

[7] Gödel, it will be recalled, maintained that 'mathematics describes a non-sensual reality, which exists independently both of the acts *and the dispositions of* the human mind and is only perceived' (Gödel, 1951: 1995, p. 323, emphasis mine).

Sense perception can therefore in some sense and in some circumstances 'force' false contents on us as being true. It does so, moreover, through the operation of deep-seated mental dispositions and not just the objective properties of the material objects involved.

Any attempt to establish an analogy between sense perception and a perception-like form of mathematical apprehension must thus address the question whether there are parallels to such phenomena as Ames' 'Distorted Room' in the case of mathematical perception. Specifically, it must consider whether (i) false as well as true contents can be forced upon us as being true, whether (ii) the phenomenon of forcedness is any surer indication of the existence and characteristics of external objects than it is of the existence and working of deep-seated mental *dispositions*, whether (iii) there is a phenomenon of 'familiarity' or 'training' that affects mathematical perception in a way similar to that in which sensory familiarity/training affects sensory judgement in the case of the 'Distorted Room',[8] whether (iv) there is a systematic connection between reliability, on the one hand, and belief-selection that favours such familiarity/training, on the other, and, finally, whether (v) there are reliable phenomenal means of distinguishing forcedness which arises from the influence of external objects from forcedness which arises from the operation of mental dispositions.

In addition, there are other problems as well. One is to discern when an attitude that registers as acceptance is indeed a case of accepting a *propositional content as true*. There are different types, degrees and modes of acceptance, and different objects of acceptance. Mere awareness of what we identify as an attitude of acceptance would therefore seem to offer little guidance concerning its exact character and the ultimate source of its compulsion. Much of what we experience as compelled acceptance is doubtless due to long, complex and largely invisible conditioning of various sorts, and we are generally far from

---

[8] The possibility of an effect of familiarity/training on belief-selection in mathematics should, I think, be taken seriously. I lack the space to explore this idea further here, but it seems to resonate with various widely advocated constraints on mathematical theorizing. Foremost among these is the so-called *Principle of Permanence*, a principle formulated and energetically defended by Peacock and others in the early and mid-nineteenth century, and a principle widely adopted by later mathematicians. A relatively recent statement endorsing (one variant of) the Principle of Permanence is the following by Courant and Robbins:

> ... the essential logical and philosophical basis for operating in an extended number domain is formalistic; that extensions have to be created by definitions which, as such, are free, but which are useless if not made in such a way that the prevailing rules and properties of the original domain are preserved in the larger domain.
>
> Courant and Robbins (1947: 1981, p. 89)

For more on the Principle of Permanence see Detlefsen (2005, pp. 273–278).

knowing its nature and origins. Frege had a particularly desultory view of the conditioning that he took to lie behind many of our affirmations.[9]

One need only use a word or symbol often enough, and the impression will be produced that this proper name stands for something; and this impression will grow so strong in the course of time that in the end hardly anybody has any doubt about the matter.

Frege (1903: 1962, Vol. II, §142)

Neither Gödel nor others since his time have adequately addressed these difficult problems. And though this should not be taken to suggest that they are irresolvable, it should motivate consideration of alternative approaches to Gödel's seeming understanding of the invention/discovery question in mathematics. It is to the consideration of such alternatives that I now turn.

Traditional alternatives

Gödel's view shares a concern with the more important traditional views concerning the discovery vs. invention issue. Broadly speaking, this is a concern for the objectivity of mathematics. A little more exactly, it is a concern to restrict certain types of subjectivity in mathematics, specifically those having to do with the possibility of legitimately using 'created' or 'invented' concepts. I'll now briefly survey some of the more interesting parts of the historical literature that touch on this issue.

Ancient views

The extent to which mathematics should be open to creation or invention was a common concern in antiquity. Looking at the ancient literature, though, we quickly notice a terminological difference. Specifically, what is currently meant by 'discovery' is quite different from what the ancients meant.

This can be seen from Cicero's (106 BC–43 BC) use of the term. He distinguished two types of methods proper to careful, systematic inquiry in any area of thought, mathematics included. Methods of the one type he called methods of 'discovering' (Cicero, 1894–1903, p. 459), methods of the other type, methods of 'deciding' (ibid.). In time this grew into the traditional division between *artis inveniendi* (arts of invention, arts of discovery or arts of investigation) and *artis iudicandi* (arts of adjudication, also sometimes called *artis demonstrandi* (arts of demonstration)).

This distinction was important in science and also in jurisprudence. In mathematics, it generally took the form of a distinction between

---

[9] In the concluding section I'll discuss two others, Bolzano and Dedekind, who also emphasized the elaborate conditioning that underlies even ordinary mathematical beliefs.

discovermental and demonstrative methods of investigation. The end of the former was the efficient development of new knowledge, albeit only perhaps of sub-optimal quality. The end of the latter was the perfection of knowledge—specifically, the perfection of the imperfect knowledge produced by the justificatively sub-optimal methods of discovery.[10]

Traditionally, then, 'discovery' and 'invention' were synonyms. They marked the first stage of a classical two-staged conception of justification which was divided into a *certificative* stage and an *argumentative* stage. The general idea behind this *Classical Scheme* can be summarized as follows:

> *Genuine scientific justification of a proposition*[11] *requires both certification and argumentation. Argumentation by itself is not enough because it can too easily be fictional. De-fictionalization of argumentation is the proper job of certification. Genuine demonstration, then, is certified argumentation, and genuine scientific knowledge requires genuine demonstration.*

This scheme figured significantly in the thinking of ancient geometers. An interesting example is Proclus (411–485) who, in his *Commentary* on Book I of the *Elements*, appealed to the Classical Scheme to explain the 'ordering' of Propositions I–IV.[12]

---

[10] It was commonly believed as well that pre-demonstrative investigation ought to be *preparative to* demonstrative investigation in the sense of making the development of demonstrations easier or more efficient. Archimedes (287 BC–212 BC), for example, recommended 'mechanical' methods to Eratosthenes as means of furthering the search for demonstrations.

> ...*I have thought fit to write out for you and explain in detail...a certain method, with which furnished you will be able to make a beginning in the investigation by mechanics of some of the problems in mathematics...this method is no less useful even for the proof of the theorems themselves. For some things first became clear to me by mechanics, though they had later to be proved geometrically owing to the fact that investigation by this method does not amount to actual proof; but it is, of course, easier to provide the proof when some knowledge of the things sought has been acquired by this method rather than to seek it with no prior knowledge.*
>
> *Archimedes (1993, pp. 221–222)*

[11] That is, justification capable of sustaining genuine scientific knowledge (*scientia, epistēmē*).

[12] Propositions I–IV are:

    I: On a given finite straight line to construct an equilateral triangle.

    II: To place at a given point (as an extremity) a straight line equal to a given straight line.

    III: Given two unequal straight lines, to cut off from the greater a straight line equal to the less.

*...our geometer [Euclid, Bk. 1] follows up these problems [Props. I–III]
with his first theorem [Prop. IV] ... For unless he had previously shown the
existence of triangles and their mode of construction, how could he discourse
about their essential properties? Suppose someone... should say: 'If two
triangles have this attribute, they will necessarily also have that.' Would it not
be easy... to meet this assertion with "Do we know whether a triangle can be
constructed at all?"... It is to forestall such objections that the author of the
Elements has given us the construction of triangles... These propositions are
rightly preliminary to the theorem...*

Proclus (1970, pp. 182–183)

Like other ancient geometers, Proclus' conception of the distinction
between certificative and argumentative methods followed the lines of the
ancient distinction between *problematic* investigations and *theorematic* investigations. The former consisted in 'the working out of problems' (Proclus, 1970,
p. 157), the latter in 'the discovery of theorems' (loc. cit.).

The aim of the former was 'to produce, bring into view, or construct what in
a sense does not exist' (loc. cit.), and accomplishment of this aim was generally
secured by 'construct[ing] a figure, or set[ting] it at a place, or apply[ing] it to
another' etc. (loc. cit.). In sum, it was generally accomplished by *construction*.

The aim of theorematic investigation, on the other hand, was 'to see, identify, and demonstrate the existence or nonexistence of an attribute' (loc. cit.). In
other words, it was an effort 'to grasp firmly and bind fast by demonstration the
attributes and inherent properties belonging to the objects that are the subject-
matter of geometry' (op. cit., pp. 157–158).

Problematic investigations were thus intended to establish the existence
of something 'new' (i.e., something whose existence had not previously been
established), while theorematic investigations were intended to establish the
properties of already existing things. Hence the need for prior certification of
the existence of the subjects of theorematic investigations.

Proclus' comment on the ordering of Propositions I–IV reflects acceptance
of such a scheme. Proposition IV is a *theorem* whose subject is triangles, lines,
equal lines, etc. Propositions I—III, on the other hand, are *problems*, which
were needed in order to establish the existence of the subject of Proposition IV.
They were thus 'rightly preliminary' (op. cit., p. 183) to it even though they
were not used in the proof for it.[13]

IV: Two triangles having two of their respective sides equal, and the angles contained by those sides equal, will also have equal bases, be equal to each other as
triangles, and have the remaining angles equal.

[13] Proposition IV is also about equal triangles, angles and equal angles. This being
so, one would naturally expect to see preliminary propositions establishing the existence of these items. Proclus did not comment on this and we can therefore only wonder
whether he regarded it as a defect in Euclid.

In saying that Propositions I–III are 'rightly preliminary' to Proposition IV, Proclus seems to have meant that they address certain broad sceptical challenges to it and/or its proof. Proposition I, for example, responds to the challenge raised by the question 'Do we know whether a triangle can be constructed at all?'. Propositions II and III, on the other hand, respond to the challenge that perhaps no straight line (or other geometrical quantity for that matter) is equal to another.[14]

Overall, then, Proclus seems to have taken the view that fabricated theorematic investigation is uncertifiable investigation or investigation that lacks a subject. To avoid it, the objects that are to form the subject-matter of a would-be theorem should be priorly shown to exist. This, Proclus believed, was (best? only?) done by giving a genetic or constructive definition of the subject—a definition which 'explains the genesis of a thing; that is, how the thing is made or done: as is this definition of a circle, viz., that it is a figure described by the motion of a right line about a fixed point' (Hutton, 1795–1796: 2000, Vol. 1, p. 362).[15]

Modern views

Mathematicians and philosophers of the modern era held similar views, as the following statement by Leibniz illustrates:

> ... the concept of circle put forward by Euclid—namely, that it is the figure described by the motion of a straight line in a plane about one fixed end— affords a real definition, for it is clear that such a figure is possible. It is useful ... to have [such] definitions... beforehand... [For] we cannot safely devise demonstrations (secure texere demonstrationes) about any concept, unless we know that it is possible; for of what is impossible or involves a contradiction (impossibilibus seu contradictionem involventibus), contradictories can also be demonstrated. This is the a priori reason why real definition is required for possibility.
>
> Leibniz (1683:1973, pp. 12–13 (294), brackets added)[16]

---

[14] More specifically, Propositions II and III provided two basic methods for producing equal lines: one when no line previously exists at a given place, the other when a longer line exists at that place (see Proclus, 1970, p. 183).

[15] Similar ideas are familiar from western jurisprudence, in particular, from its general adoption of the principle of *corpus delicti*. This requires evidence not only for justified *conviction* but for justified *trial* as well. The idea is that for a trial to be justified there must be evidence both of the 'existence' of a crime and of the identity of its perpetrator. In the case of murder, this has typically taken the form of evidence of death as the result of an act, and evidence of the criminal agency of an identified person as its means. Other types of crimes call for other types of evidence of course. All legitimate trial, however, requires evidence of both the existence of a crime and the identity of the criminal.

[16] The number in parentheses is the page number of the Latin text in Leibniz (1978, Vol. 7). For a similar statement see Leibniz (1989, XXIV).

Leibniz thus held that genetic or constructive definitions are valuable because they make known a priori the possibility of the concepts they define (see Leibniz, 1764: 1916, Bk. II, ch. II, §18). This in turn, he suggested, makes demonstration 'safe' in a certain way.

He made a confusing variety of claims concerning impossible concepts and their existence, however. Sometimes he suggested that there are impossible concepts (i.e., concepts that imply a contradiction).

> *A concept is either suitable or unsuitable. A suitable concept is one that is established to be possible, or not to imply a contradiction.*
>
> *Leibniz (1903, p. 513)*

Other times he suggested that impossible concepts cannot exist. Thus, for example, he claimed that 'we cannot have any idea of the impossible' (Leibniz, 1978, Vol. 4, p. 424), and he also maintained that 'what actually exists cannot fail to be possible' (Leibniz, 1978, Vol. 7, p. 214).[17]

My main concern here, though, is with the suggestion that there is something unsafe about demonstration concerning A when it is not accompanied by knowledge of the possibility of A. Specifically, I'm concerned with the reasons Leibniz may have had for believing this.

The danger he mentioned was a possibility of contradiction—'of what is impossible...contradictories can be demonstrated' (loc. cit.). Taken as we would take it today, though, this does not seem right. The principal type of theorem in Leibniz' time was a proposition of the form 'All A are B'. The contradictory of this would be (a sentence equivalent to) a sentence of the form 'Some A are not B', which requires the existence of A's.

When A is impossible, though, A's cannot exist. Hence, one who, like Leibniz, maintained that the actual must be possible, would not have believed it possible to demonstrate 'Some A are not B' for impossible A. Consequently, he would have had no reason to regard demonstration of 'Some A are not B' as posing a threat to demonstration of 'All A are B' when A is impossible.

More likely, Leibniz was thinking of demonstration of 'No A are B' rather than demonstration of 'Some A are not B' as the imagined counter-demonstration to demonstration of 'All A are B'. Supposing this to be right, though, we are left with the problem of saying what it is about such a situation that would make it 'unsafe'. Whatever else Leibniz may have thought the danger to be, it would not be one of directly demonstrating a false statement. That this is so follows from elementary (classical) logical facts. When A is impossible, there are necessarily no A's, and when this is so 'All A are B' and 'No A are B' are both true.[18]

---

[17] See Mates (1986, pp. 66–68) for a brief but useful discussion of these puzzling matters.

[18] This supposes of course that A may still be a genuine concept even when it is impossible. There were, of course, many who challenged this, on which more later.

But if it is not the falsehood against which real definition of A protects us, from what does it protect us? Leibniz was not explicit on this point.[19] He may have seen the threat as a threat of fictionalization—that is, the threat of engaging in investigation that has no subject and hence, in the final analysis, is about nothing and produces knowledge of nothing. This seems to have been the classical view and was essentially the view Proclus was urging. Or so I would argue. Others of Leibniz' time, and of earlier and later times too, were more definite, and they generally gave one of two views of the security issue, one broadly practical, the other broadly theoretical in character. I'll presently describe and discuss each of these. For the moment, though, the main points to bear in mind are that (i) there was and perhaps still is a tradition in the history of mathematics of linking the reliable with the broadly experiential, at least to the extent that construction was taken to be a medium of experience, but that (ii) the content of such experience was not necessarily that of the proposition ultimately being proved or justified.

Real definition as a practical concern

The practical defense of real definition did not challenge the common idea that the reality or possibility of a concept is constituted by its consistency. Its claim was rather that the only practical means of establishing the consistency of a concept was to exhibit an instance of it, that is, to give a real definition of it.[20]

The efficacy of real definitions as a means of ensuring consistency was commonly seen as the reason why classical geometry, in which real definitions predominated, was so much less problematic than algebra, in which they did not. The following remark by Playfair is a typical eighteenth-century expression of this thinking:

> *The propositions of geometry have never given rise to controversy, nor needed the support of metaphysical discussions. In algebra, on the other hand, the doctrine of negative quantity and its consequences have often perplexed the*

---

[19] Leibniz' example of a real definition was Euclid's definition of a circle—namely, that it is 'the figure described by the motion of a straight line in a plane about one fixed end' (loc. cit.). This was in keeping with the general understanding of real definition in the seventeenth, eighteenth and nineteenth centuries. Hutton, for example, gave the following characterization in his *Dictionary*:

> Real Definition ... *explain[s] the genesis of a thing; that is, how the thing is made or done: as is this definition of a circle, viz., that it is a figure described by the motion of a right line about a fixed point.*

> *Hutton (1795–1796: 2000, Vol. 1)*

[20] By the 'consistency' of a concept is meant the consistency of the statements that are taken to describe it or to apply to it.

analyst, and involved him in the most intricate disputations. The cause of this diversity must no doubt be sought for in the different modes which they employ to express our ideas. In geometry every magnitude is represented by one of the same kind; lines are represented by lines, and angles by an angle, the genus is always signified by the individual, and a general idea by one of the particulars which fall under it. By this means all contradiction is avoided, and the geometry is never permitted to reason about the relations of things which do not exist, or cannot be exhibited.

Playfair (1778, pp. 318–319)

The practical identification of demonstration of consistency with production of an instance became, if anything, even more definite in the nineteenth century. The following statement by John Herschel expressed the common attitude:

The test of truth by its application to particulars being laid aside, nothing remains but its self-consistency to guide us in its recognition. But this in axiomatic propositions amounts to no test at all...Their [the axioms'] mutual compatibility as fundamental elements of the same body of truth, can only be shown by experience—by the observed fact of their coexistence as literal truths in a particular case produced.

Herschel (1841, p. 220, brackets added)

Frege too pressed the point in his disagreement with the creativists (e.g., Hankel and Hilbert). He thus responded to Hankel's identification of consistency and existence by observing;

Strictly, of course, we can only establish that a concept is free from contradiction by first producing something that falls under it. The converse inference is a fallacy, and one into which Hankel falls.

Frege (1884:1968, §95)

Similarly, in his correspondence with Hilbert concerning his (Hilbert's) seeming reliance on the possibility of proving the consistency of a concept without providing a verifying instance of it, Frege wrote:

What means have we of demonstrating that certain properties, requirements (or whatever else one wants to call them) do not contradict one another? The only means I know is this: to point to an object that has all those properties, to give a case where all those requirements are satisfied. It does not seem possible to demonstrate the lack of contradiction in any other way.

Frege to Hilbert, January 6, 1900 (Gabriel et al., eds., 1980, p. 43)

Eventually, frustrated by Hilbert's intransigency, he challenged him to explain how it is possible to prove the consistency of a concept other than by identifying a verifiable instance of it.

I believe I can deduce, from some places in your lectures, that my arguments failed to convince you, which makes me all the more anxious to find out your counter-arguments. It seems to me that you believe yourself to be in possession of a principle for proving lack of contradiction which is essentially different from the one I formulated in my last letter and which, if I remember right, is the only one you apply in your Foundations of Geometry. *If you were right in this, it could be of immense importance, though I do not believe in it as yet, but suspect that such a principle can be reduced to the one I formulated and that it cannot therefore have a wider scope than mine. It would help to clear up matters if in your reply to my last letter—and I am still hoping for a reply—you could formulate such a principle precisely and perhaps elucidate its application by an example.*

Frege to Hilbert, September 16, 1900 (Gabriel et al., eds., 1980, p. 49)

Frege's reference to Hilbert's *Grundlagen der Geometrie* was to his proof of the consistency of his system of geometry by interpretation in the real numbers (see Hilbert, 1899, §9). He was right to note that this was the only proof of consistency Hilbert offered at the time. It would be another twenty years before his proof-theoretic conception of consistency was even outlined.[21]

That production of a verifying instance was the only way to establish the consistency of concept was thus a prevalent belief of the seventeenth, eighteenth and nineteenth centuries. It is also one which allows us to explain why Leibniz and others should have seen demonstration of 'No A are B' as posing a threat to demonstration of 'All A are B'. The explanation is as follows. If both 'All A are B' and 'No A are B' are provable, then, supposing that what is provable is true, there will be no A's. If, however, there are no A's, there will in practice be no way of establishing the consistency of all that is eventually asserted of A (including, *ex hypothesi*, 'All A are B'). The only practical way of doing this is to produce an *instance* of A and to verify of that instance that it has the properties attributed to it by the assertions that have it as their subject-concept. In the end, then, we need instantiation of A because we eventually will need to prove the consistency of the statements made about A, and, as a matter of practical necessity, this can only be done by producing an A.

The supposed upshot of this reasoning for the question whether mathematics is created or discovered is this: thoroughgoing creativism regarding concepts is unsustainable. The reason is that every theory must be consistent, and, in the final analysis, this can only be shown by identifying a witnessing instance. Such witnesses, Arnauld observed, can only be *discovered*, not *created*.

---

[21] In my view it is unclear how closely a model-construction proof like Hilbert's resembles the type of constructive or concrete instantiation that Leibniz and Frege seemed to take real definition to provide. The model Hilbert offered for geometry was not one which persuades through visualization or experience. It's far more abstract than Euclid's real definition of a circle and it presumes an ability to make abstract judgements. That model-construction of the type Hilbert used is really by *intuition*, or is more *discursive* in character, and, so itself in need of intuitive vindication seems open to doubt.

...nominal definitions are arbitrary but real definitions are not. Since any sound is ...capable of expressing any idea whatever, I am permitted for my own use to choose a certain sound to express ...one idea. But the case is completely different with real definitions, for the relations which obtain among ideas are independent of man's will.

Arnaold (1964, p. 83)

The question of the proper roles of nominal and real definition were thus at the heart of the traditional discussion of the discovery vs. creation issue. It remained a central element well into the twentieth century. Understanding this, I believe, should help us both to deepen and to broaden our own perspective on the issue. More particularly, it should lead us to an appreciation of what more may lie behind the issue, and what more may be at stake in its resolution than arguments like Gödel's suggest.

Real definition as a theoretical concern

The theoretical defence of real definition is based on a view of how we acquire concepts. This is a broadly empiricist or experientialist view which sees abstraction from experience as the fundamental means by which concepts are developed. It therefore classifies concepts as *real* or not according to whether there is a plausible account of how they might have arisen, and whether an abstractive path from experience is also specified. Concepts which arise in this way are legitimate or real, those which do not are not.

This view has had many advocates among mathematicians and philosophers from the modern era onward. John Leslie's popular early nineteenth century geometry text, for example, taught that 'Geometry is...founded on external observation, but such observation is so familiar and obvious that the primary notions which it furnishes might seem intuitive, and have been regarded as innate' (Leslie, 1809, p. 2). And again 'Geometry, like the other sciences which are not concerned about the operations of the mind, rests ultimately on external observation' (op. cit., p. 453).

A common synonym for 'real' among broadly empiricist thinkers of the eighteenth and nineteenth centuries was the term 'given', a term Leslie characterized as follows in the sequel to the text cited above:

Quantities *are said to be* given, *which are either exhibited, or may be found.*

Leslie (1821, p. 4)

Overall then, the theoretical defence of real definition in geometry was based on two ideas. The first was that geometrical concepts are derived from observation of external objects by a process(es) of abstraction. The second was that the contents of such observation are in an important sense *given* rather than created. Real concepts are thus representations that begin with a content that is *given*, not one that is *created* or fabricated. They then arise by a process of abstraction from this experience, a process which only subtracts from and does not add to the original experienced contents.

This view of real concepts was sometimes extended to a general view of concepts by nineteenth-century philosophers. A striking example was Schopenhauer, who offered the following memorable account of the nature of concepts and the conditions under which they exist.

> ... *concepts derive their content from the intuitive realm and therefore the entire structure of the world of thought rests upon intuitions. We must therefore be able to go back from every concept, even if indirectly through intermediate concepts, to the intuitions from which it is itself abstracted ... That is to say, we must be able to support it with intuitions which stand to the abstractions in the relation of examples...*
>
> *These intuitions... afford the real content* (realen Gehalt) *of all our thought, and whenever they are wanting we have not had concepts but mere words in our heads* (blosse Worte im Kopfe). *In this respect our intellect is like a bank which holds notes* (Zettelbank), *which, if it is to be sound, must have cash in its safe, so as to be able to meet all the notes it has issued, in case of demand. The intuitions are the cash and the concepts the notes.*
>
> *Schopenhaver (1859: 1966, Book 1, ch. 7, vol. 2)*[22]

Views of this general type were not, however, confined to philosophers or to the nineteenth century. Indeed, no lesser nor less recent a mathematician than Hermann Weyl used the image of notes of deposit to describe the difference between genuine concepts and propositions, on the one hand, and purely symbolic devices, on the other. He used the existence claim in mathematics as his chief example, and he likened it to paper money—that is, money which in itself has no value and whose only true value is that of a real commodity that backs it.

> *An existential proposition* (Existentialsatz)—*something like 'there is an even number'—is not at all a judgment in the genuine sense of an assertion of a fact. Existential states of affairs* (Existential-Sachverhalte) *are an empty invention* (Erfindung) *of the logician. '2 is an even number': that is a real* (wirkliches), *a factual expression of a given judgment. 'There is an even number' is only a judgment-abstract* (Urteilsabstrakt) *obtained from this judgment. Just as I take knowledge* (Erkenntnis) *to be a valuable* (wertvollen) *holding* (Schatz), *so I regard the judgment-abstract as paper* (Papier), *which somehow indicates* (anzeigt) *a holding without disclosing* (verraten) *where it is. Its only value can lie in its ability to get me to search for the holding. The paper is worthless so long as it is not realized* (realisiert) *by a real* (wirkliches) *judgment such as '2 is an even number' that stands behind it.*
>
> *Weyl (1921, p. 54)*

---

[22] See Schopenhauer (1911, p. 76) for German version. The first edition was published in 1819, the second in 1844, a third in 1859. The passage quoted here is in all these editions.

In all the above we see the same basic idea at work, namely, that those representations we call concepts exist only to the extent that they stand in a relationship of abstractability to the contents of experiences that are given and not created.

This offers a relatively clear explanation of how it is that proof of 'No A are B' would trivialize or devalue proof of 'All A are B'. If both 'All A are B' and 'No A are B' expressed propositional contents and were true, there would be no A's. Hence there would be no experiential content—that is, no experienced instance of A—from which A could be abstracted. The expression 'A' would therefore have no content and would not express a concept. As a result, 'All A are B' would not express a proposition and could not therefore be a content of genuine judgement. As a result, a demonstration of 'All A are B' would not yield knowledge that all A are B, and, so, would not achieve its intended purpose.

But though this reasoning is clear, it is also unconvincing, and for at least two reasons. The first is its foreclosure of the possibility of *uninstantiated* concepts. This is troubling since there seem to be genuine concepts (or at least concept-like representations) that not only are not instantiated but, indeed, are not capable of being instantiated. This at any rate for composite concepts (e.g., that of a map requiring five colors for its colouring). Whether there are concepts that can properly be regarded as primitive and are not instantiable is a more difficult question to answer. I know of no good argument that there could not be such contents, however.

The second reason why the Schopenhauer–Weyl argument is unconvincing is its seeming blindness to the possibility that the type of experience capable of sustaining derivation of a concept by abstraction might itself presuppose the availability of concepts. The Schopenhauer–Weyl argument supposes that we could have the types of contents needed to begin a process of abstraction without having access to concepts that are not based on such acquisition. To be convincing, though, a closer analysis of contents and abstraction would have to be offered, and, with it, an argument that there are at least some types of rudimentary contents we can grasp without prior application of concepts. In addition, reasons would have to be given for the claim that there is a plausible path of acquisition from such contents to mathematical contents via abstraction.

Until such gaps are filled, the theoretical defence cannot readily be accepted as a general account of the existence or acquirability of concepts. But though it has shortcomings as a general theory of mathematical concepts, the theoretical defence might yet provide a basis for a worthwhile distinction between discovery and creation in mathematics. The idea would be that a real concept is an instantiable concept. The existence of non-instantiable concepts would not be denied. Neither would their use be forbidden. Their use, however, would be controlled by an appropriate form of consistency requirement—specifically, a consistency requirement the demonstrated satisfaction of which would not necessitate instantiation of the concept(s) involved.

Discussion

Both of the traditional defences of real definition discussed above offer possibilities of significant distinctions between discovery and creation in mathematics. Both also diverge from Gödel's phenomenological argument in that they see the distinction between discovery and creation as wanted for more than merely a phenomenologically adequate accounting of whatever experience we may have of truth being 'forced' on us. It is not necessary that they deny that there is a point behind Gödel's phenomenological reasoning. Neither, though, do they see it as indicating the primary role for experience in the development of mathematical knowledge.

Both defences take the view that producing an instance of a concept is fundamentally a matter of discovery rather than of invention. Moreover, they both see discovery as discovery of *reality* where reality guarantees consistency or possibility. Creation, on the other hand, does not. In order for real definition to fulfill its intended justificative purpose, then, it is necessary that it be seen as a case of discovery rather than creation. Otherwise, the mathematician would possibly be faced with an endless regress of proper justificative demands, and this is surely something she should want to avoid.

I mentioned above that the practical defence of real definition seems to underestimate the variety of forms a consistency constraint on the use of created concepts might take. This was due to the common belief prior to Hilbert's development of proof theory that the only way to prove the consistency of a concept was to find a recognizable instance. Hilbert's proof-theory provided a syntactical alternative to this.

On the other hand, Gödel's incompleteness theorems (particularly his second theorem) suggest that there are limits to the applicability of the Hilbertian alternative. This may mean that, practically speaking, we will often have to rely on production of instances or construction of models to prove consistency. In the end, then, Frege may have been right concerning the practical possibilities of proving consistency.[23] Therefore, in the final analysis, it is difficult to determine how extensive a role for discovery the practical defence is capable of supporting.

Conclusion

I would like to close by considering one final question concerning Gödel's proposed use of the phenomenon of given-ness or forcedness in mathematics. This time, though, my concern is not whether we can reliably detect it and, if so, with what degrees of confidence and discernment. Rather, it is with what a proper response to experiences of 'given-ness' or 'forcedness' might or ought to be in the first place, supposing that we do in fact have them. I was led to this

---

[23] There are of course serious questions concerning the similarity of model-construction to the kind of exhibition of an instantiating instance supposed to typify real definition. I lack the space to go into these here, however.

by Bolzano's trenchant questioning of the role and character of real definition in mathematics.[24]

As is well-known, Bolzano's proof of the intermediate value theorem was largely the product of his concern that the methods used to prove a theorem be of a level of generality appropriate to that of the theorem proved. In his view, a general theorem about quantity ought to appeal only to general laws of quantity, and never to laws pertaining only to *some particular type of* quantity.

This led him to reject previous proofs of the intermediate value theorem on account of the appeals they made to specifically geometrical quantities. It also led him to reject the usual proofs of geometrical theorems, many of which he believed to follow from general laws of quantity. A more accurate picture of the true grounds of such theorems would be given by proofs whose premises were themselves general laws of quantity.

> ... *in* Euclidean *geometry no spatial object is accepted as real unless its construction has first been demonstrated by means of* plane, circle *and* straight line. *This restriction betrays its* empirical *origin clearly enough.* Board, compass, *and* ruler *are...the simplest instruments which were needed initially for drawing. However, considered in themselves the* straight line, the circle, *and also the* plane *are such compound objects that their possibility cannot be accepted in any way as a* postulate... *For example, the proposition that between every two points lies a* mid-point *is far simpler than the proposition that between every two points a* straight line *can be drawn. Nevertheless,* Euclid *proves the former from the latter and several others. It is sufficient for the* theoretical *exposition of mathematics...that one proves the* possibility *of every conceptual connection which is put forward. How, and in what way, an* object *analogous to the concept can be produced in* reality *belongs to* practical mathematics.[25]

> *Bolzano (1810: 2004, §37)*

In my view, part of what Bolzano was doing here was to point out dangers of following the leadings of forcedness. Too much trust may lead to such things as proving the more elementary (though perhaps less forced) from the less elementary (but more forced).

This raises a related question concerning Gödel's views. He said that it is the *axioms* of set theory that force themselves on us as being true. At the

---

[24] Dedekind raised similar concerns in arithmetic. In the first paragraph of *Was sind und was sollen die Zahlen?* he thus warned against following the leadings of inner intuition (*innere Anschauung*), and said that evident truths are quite often those most in need of proof. This indeed became the leitmotif of this foundational work in arithmetic, the central principle of which was that nothing at all capable of proof should be accepted without proof, regardless of the degree of its evidentness. See Dedekind (1888), preface to the first edition.

[25] In this sentence, the reality of which Bolzano speaks is empirical or sensible reality. He thus demotes the type of real definition found in Euclid from scientific mathematics to what was then regarded as applied mathematics.

same time, he likened forcedness in mathematics to sensory perception. This would lead us to think that the axioms of set theory are to the larger domain of mathematical truth what sense perceptions are to the larger domain of natural scientific truth.

If this is correct, though, then, pursuing the analogy between mathematics and natural science, the propositions of set theory that force themselves on us as being true are not likely to be the basic laws of mathematics. That is, they should not properly be taken to be axioms. Rather, they should be seen as phenomena whose proper explanation requires deeper and more basic laws, a *theoretical* mathematics if you will.[26] It is not easy, however, to think what these deeper laws might be. In addition, the analogy between the axioms of set theory and sensory judgements does not square well with what Gödel himself said about the relationship of set theory to number theory—namely, that the former can be at least partially justified by its ability to simplify proofs of theorems of the latter.

Overall, then, there are serious problems concerning the analogy between sensory experience and the type of experience by which Gödel supposed various truths of mathematics to force themselves on us as true. If forcedness is symptomatic of experience, then the mathematical propositions forced on us as true ought to be seen as the experiential part of mathematics. Pursuing Gödel's analogy we would then be led to consider the possibility of an observation/theory divide in mathematics, similar to that which we accept in natural science. Bolzano believed in something like this. Specifically, he believed in objective differences of basicness between mathematical propositions. Moreover, he took it as the principal duty of the mathematician to reveal this ordering of relative basicness.

Bolzano thus raised important questions concerning given-ness and its possible significance to mathematical epistemology. I've indicated some of the challenges these questions raise for Gödel's attempt to introduce a type of experience into mathematical epistemology. These should warn us against equating degree of given-ness or forcedness with degree of basicness. They do not, so far as I can see, speak with similar force against the traditional understanding of the discovery/creation issue in mathematics—namely, that which centers on the use of so-called 'real' definition in mathematics and the protection it may provide against certain of the grosser forms of subjectivity.

---

[26] Russell and others, of course, expressly adopted this 'regressive' conception of the axioms of set theory.

Comment on Michael Detlefsen's 'Discovery, invention and realism'

John Polkinghorne

━━━━━━━━◦◦◦◦━━━━━━━━

Michael Detlefsen gives us a careful discussion of the claim made by Kurt Gödel that we have experience of being confronted by mathematical objects in a non-negotiable manner, similar to the confrontation with physical objects that persuades us of the independent reality of the physical world. He is surely right to assert that this claim, if it can be substantiated, provides the best ground for the defence of the conviction held by many mathematicians that their reasoning is a matter of discovery rather than the mere invention of pleasing intellectual puzzles.

In probing this analogy, I think it is necessary to take fully into account the subtlety of our encounter with the physical world. Defenders of realist claims for modern physics, of whom I am one (Polkinghorne, 1996, ch. 2), are not simply appealing to the fact that we bump into large objects. The quantum world, for example, is something much too veiled and elusive in its character to be treated as if it were simply naively objective. Nevertheless, physicists are convinced of the reality of entities such as electrons or quarks, refusing to treat them as if they were just imaginative and imaginary devices to enable calculations to be made. I believe that the defence of physical realism ultimately depends upon the intelligibility that these entities enable us to attain. We take electrons and quarks with ontological seriousness because their existence explains great swathes of more directly accessible phenomena. The power of group theory to illuminate the structures of symmetrical patterning (du Sautoy, 2008) seems to offer an analogy in favour of taking the noetic reality of finite groups seriously.

Another argument in favour of the reality of the physical world is its character of quite often proving surprising, with properties contrary to prior 'reasonable' expectation. Quantum theory is the prime exemplar of this. This resistance to prior expectation is persuasive that we are encountering an independent reality standing over against us. I suppose that the nineteenth-century discovery of non-Euclidean geometries would be a mathematical analogy.

Detlefsen recalls Kant's distinction between intuitions, which come to us with the inexorable character of giveness, and inventions, which he believed can be created and manipulated at human will. It seems to me that there is some connection here with an experience testified to by mathematicians, of the coming into consciousness of a deep mathematical theory 'fully-formed', so to speak. In my own chapter I referred to a famous anecdote of how Henri Poincaré, after months of struggle with Fuchsian functions, found that the complete theory sprang into his mind as an unbidden gift just as he was setting off on holiday. The non-negotiable character of mathematical ideas seems illustrated by the fact that a mathematician considering a particular axiomatized system can 'see' the truth of the Gödelian sentence, despite its being formally unprovable in that system.

Einstein once said that the basis of fundamental physics has to be 'freely invented'. He certainly was not subscribing to a postmodern idea of the subjective creation of merely pleasing notions. Einstein was too uncompromisingly objective in his thinking for that. Rather, he was pointing to the role of the leap of creative imagination that enabled him, for example, to write down the beautiful equations of general relativity, even if they then had to be verified by comparison with phenomena. Great ideas in mathematics are similarly grasped which display the property of being 'deep'—that is, an apparently simple formulation proves to have extensive and unexpected consequences, in analogy to the range of prediction of a successful physical theory when it is found to imply consequences not even empirically known when the theory was discovered. Think of the astonishing fruitfulness of the idea of complex numbers.

Defences of realist interpretations in both physics and mathematics have to be subtle and delicate and it seems to me that the two disciplines are cousins under the skin in this respect.

8
Mathematics and objectivity

Stewart Shapiro

I wish to explore, in a tentative and general way, the extent to which mathematics is objective. As is typical in philosophy, part of our question is to try to get clear on the meaning of the terms in the question. I take it for granted that we know what mathematics is, or at least that we know it when we see it, borderline cases aside. But what of 'objectivity'? Intuitively, to be objective is to be independent of human judgements, conventions, forms of life and the like. But what of this notion of 'independence'?

The view that mathematics is not objective—that mathematical truths are somehow tied to the nature of human cognition, conventions or whatever—is not uncommon among philosophers. Immanuel Kant took mathematics to flow from 'pure intuition', the form of our faculty of perceiving the world spatially and temporally. Thus, mathematics is directly connected to human abilities, and so lacks objectivity—at least in some sense of that term. The traditional intuitionists, L. E. J. Brouwer and Arend Heyting, follow suit. Heyting (1931: 1983, p. 52) wrote:

> The intuitionist mathematician proposes to do mathematics as a natural function of his intellect, as a free, vital activity of thought. For him, mathematics is a production of the human mind... [W]e do not attribute an existence independent of our thought, i.e., a transcendental existence, to... mathematical objects... [M]athematical objects are by their very nature dependent on human thought.

A prominent, contemporary philosopher, Terrence Horgan (1994, p. 99) adopts a convention, due to Hilary Putnam, that words written in small capitals are meant to be 'about denizens of the *mind-independent, discourse-independent,* world'. It would beg the present question if this supposed mind-independent, discourse-independent world was thoroughly material, and neither Horgan nor Putnam assumes it is.

Horgan (1994, pp. 100–101) claims that

> there is a whole spectrum of ways that a sentence's correct assertibility can depend upon THE WORLD. At one end of the spectrum are sentences governed by...norms...[such that a sentence in those discourses] is true only if some unique constituent of THE WORLD answers to each of its singular terms...At the other end of the spectrum are sentences whose governing assertibility norms...are such that those sentences are sanctioned as correctly assertible by the norms alone, independently of how things are with THE WORLD.

He then adds, parenthetically that 'sentences of pure mathematics are plausible candidates for' this second status, of being assertible or not 'independently of how things are with THE WORLD'. So Horgan takes it as 'plausible', without further comment, that pure mathematics lacks any objectivity.

Galileo Galilei's 1623 book *Il Saggiatore* (*The Assayer*) contains the much quoted passage:

> Philosophy is written in this grand book—I mean the Universe—which stands continually open to our gaze, but it cannot be understood unless one first learns to comprehend the language and interpret the characters in which it is written. It is written in the language of mathematics, and its characters are triangles, circles and other geometrical figures, without which it is humanly impossible to understand a single word of it; without these, one is wandering in a dark labyrinth.

Today, we are in an even better position to appreciate the deep truth underlying this than they were in the early decades of the seventeenth century. There is hardly a branch of natural or social science that does not have substantial mathematical prerequisites. One cannot get beyond the first few pages of a textbook without considerable mathematical facility. Of course, this is *applied* mathematics, while Horgan comments on pure mathematics. But the difference between the more theoretical parts of some of the sciences and pure mathematics is not all that clear, at least to me.

Views, like Horgan's, that deny the objectivity of even pure mathematics make a mystery of Galileo's observation—in a pejorative sense of 'mystery'. If mathematics is governed by no more than human conventions of assertibility, as Horgan suggests, then why is mathematics so important for science? There must be something about the WORLD, as it is in itself, independent of human concerns, judgements, etc., that puts mathematics at the centre of just about all of our attempts to understand it—even if they are *our* attempts to understand the universe.

Galileo speaks of the *language* of the universe. Theology aside, I presume that this is a metaphor. More literally, the claim is that one must invoke mathematics in order to fully or properly *understand* the universe, at least in a scientific manner. Does it follow that mathematical assertions themselves are objective? It is probably also true that one cannot understand the universe at any sophisticated depth without understanding a language. Does it follow that languages themselves are directly tied to the nature of the world, objectively, independent of human interests, concerns, judgements, and the like? In any case, there must be *something* about the non-linguistic world that makes languages what they are, and make them effective in communication for us humans. Similarly, there must be something about the non-mathematical world that makes mathematics effective—indeed essential—for us to understand just about anything.

To be fair, Heyting and Horgan are surely correct that mathematics and, for that matter, language and science, are human *activities*, and the pursuit and results of those activities are shaped by human concerns and interests. It is a truism that theories and explanations, in both mathematics and science, are due to both the nature of the non-human world and the nature of human knowers and understanders. As John Burgess and Gideon Rosen (1997, p. 240) put it, 'our theories of life and matter and number are to a significant degree shaped by our character, and in particular by our history and our society and our culture.' Of course, this is not to say that the *world itself* or, as Horgan or Putnam might put it, THE WORLD ITSELF, is somehow shaped by 'our character'. The Burgess–Rosen observation is that it is our *theories* of the world that are so shaped. The question before us is the extent to which the truths of mathematics are due to the way the non-human world is, and the extent to which these truths are due to the way the human mathematicians are.

Several competing philosophical traditions have it that there is no way to sharply separate the 'human' and the 'world' contributions to our theorizing. As Protagoras (supposedly) said, 'man is the measure of all things'. On some versions of idealism, not to mention some postmodern views, the world itself has a human character. The world is shaped by our judgements, observations, etc. And so, it would seem, there just is no WORLD, in Horgan's and Putnam's sense. A less extreme position is Kant's doctrine that the *ding an sich*, (or *DING AN SICH*) is inaccessible to human inquiry. We approach the world through our own categories, concepts, and intuitions. We cannot get beyond those, to the world (or THE WORLD) as it is, independently of said categories, concepts and intuitions.

On the contemporary scene, a widely held view, championed by W. V. O. Quine, Hilary Putnam, Donald Davidson, and Burgess, has it that, to use a crude phrase, there simply is no God's eye view to be had, no perspective from which we can compare our theories of the world to the WORLD itself, to figure out which are the 'human' parts of our successful theories and which are the WORLD parts.

This Kant–Quine orientation may suggest that there simply is no objectivity to be had, or at least no objectivity that we can detect. If this is right, then there simply is no answering the question of this paper. For what it is worth, I would resist this, despite sympathy with the Kant–Quine orientation. There may not be such a thing as *complete* objectivity—whatever that would be— but it still seems that there is an interesting and important notion of objectivity to be clarified and deployed. There seems to be an important difference—a difference in kind—between statements like 'pure water contains two hydrogen atoms for each oxygen atom' and statements like 'Broccoli is disgusting' and 'Manchester United is evil'. Our question concerns which side of this divide contains mathematics.

Crispin Wright's *Truth and Objectivity* (1992) contains an account of objectivity that is more comprehensive than any other that I know of, providing a wealth of detailed insight into the underlying concepts. Wright does not approach the matter through metaphysical inquiry into the fabric of REALITY, wondering whether THE WORLD contains things like moral properties or numbers. He focuses instead on the nature of various *discourses*, and the role that these play in our overall intellectual and social lives. That is, Wright tries to clarify what it is for us—the community of language users—to treat a stretch of discourse as objective, as we attempt to negotiate and understand this world we find ourselves occupying.

As Wright sees things, objectivity is not a univocal notion. There are different notions or axes of objectivity, and a given chunk of discourse can exhibit some of these and not others. The axes are labelled 'epistemic constraint', 'cognitive command', 'the Euthyphro contrast' and 'the width of cosmological role'. In a previous paper, Shapiro (2007), I argue that with the possible exception of some troubling matters near the foundation, mathematics easily passes all four tests. Mathematics is epistemically unconstrained: there are unknowable truths. The Galileo observation attests to the extremely wide cosmological role of mathematics: it figures in all sorts of explanations, most of which are explanations of non-mathematical matters. Mathematics falls on the Socrates side of the Euthyphro contrast—it is not response-dependent—and mathematics easily satisfies cognitive command. In short, mathematics is objective, if anything is.

On the other hand, the possible exceptions—the foundational matters—loom large, since they go the heart of the Kant–Quine matter noted just above. Here I want to revisit one of the axes, cognitive command, since it bears out this theme, even more than I anticipated in the other paper.

According to Wright, cognitive command figures as an axis of objectivity only if the discourse is epistemically constrained. So I will pause for a brief sketch of that primary axis of objectivity.

Epistemic constraint is an articulation of Michael Dummett's notion of anti-realism. According to one of Wright's articulations of this axis (1992, p. 75), a discourse is epistemically constrained if, for each sentence P in the discourse,

$$P \leftrightarrow P \text{ may be known.}$$

In other words, a discourse exhibits epistemic constraint if it contains no unknowable truths.

It seems to follow from the very meaning of the word 'objective' that if epistemic constraint fails for a given area of discourse—if there are propositions in that area whose truth cannot become known—then that discourse can only have a realist, objective interpretation:

To conceive that our understanding of statements in a certain discourse is fixed... by assigning them conditions of potentially evidence-transcendent truth is to grant that, if the world co-operates, the truth or falsity of any such statement may be settled beyond our ken. So... we are forced to recognise a distinction between the kind of state of affairs which makes such a statement acceptable, in light of whatever standards inform our practice of the discourse to which it belongs, and what makes it actually true. The truth of such a statement is bestowed on it independently of any standard we do or can apply... Realism in Dummett's sense is thus one way of laying the essential groundwork for the idea that our thought aspires to reflect a reality whose character is entirely independent of us and our cognitive operations.

Wright (1992, p. 4)

In other words, if epistemic constraint fails for a given discourse, then it is objective, and that is the end of the story. The other axes of objectivity— cognitive command, cosmological role and the Euthyphro contrast—are irrelevant; they do not apply. On the other hand, if a discourse *is* epistemically constrained—if all truths are knowable there—then the other axes track important aspects of objectivity. Or so Wright argues. For present purposes, then, let us just assume that, in mathematics, all truths are knowable, in some relevant sense of 'knowable', and turn to a brief characterization of cognitive command.

Assume that a given area of discourse serves to describe mind-independent features of a mind-independent world, understood intuitively. Suppose that two people disagree about something in that area. It follows that at least one of them has *misrepresented* reality, and so something went wrong in his or her appraisal of the matter. Suppose, for example, that two people are arguing whether there are seven, as opposed to eight, spruce trees in a given yard. Assuming that there is no vagueness concerning what counts as a spruce tree and no vagueness concerning the boundaries of the yard,[1] it follows that at least one of the disputants has made a mistake: she either did not look carefully enough, her eyesight was faulty, she did not know what a spruce tree is, she made a mistaken inference, she counted wrong, or something else along those lines. That is, the very fact that there is a dispute suggests that one of the disputants has what may be called a *cognitive shortcoming* (even if it is not always easy to figure out which one of them it is).

In contrast, two people can disagree over the cuteness of a given baby or the humour in a given story without either of them having a cognitive shortcoming. One of them may have a warped sense of taste or humour, or perhaps no sense of taste or humour, but there need be nothing wrong with his *cognitive* faculties. He can perceive and reason as well as anybody.

The present axis of objectivity turns on this distinction, on whether there can be blameless disagreement. Wright (1992, p. 92) writes that

---

[1] We will return to vagueness later.

A discourse exhibits Cognitive Command if and only if it is a priori that differences of opinion arising within it can be satisfactorily explained only in terms of "divergent input", that is, the disputants working on the basis of different information (and hence guilty of ignorance or error...), or "unsuitable conditions" (resulting in inattention or distraction and so in inferential error, or oversight of data, and so on), or "malfunction" (for example, prejudicial assessment of data... or dogma, or failings in other categories...

Intuitively, cognitive command holds for discourse about spruce trees and it fails for discourse about the cuteness of babies and the humour of stories.

What of mathematics? Wright takes it as obvious that cognitive command holds for simple calculations (p. 148). Suppose, for example, that two people differ on the product of two four-digit numbers, after doing the calculation by hand. Clearly, at least one of them made a mistake. He forgot the relevant multiplication table (or did not look at it accurately), or else he got the columns mixed up, or had a lapse in memory or concentration. Any of these clearly qualifies as a cognitive shortcoming. To fully sanction Wright's conclusion here, we would have to deal with Wittgensteinian issues concerning rule-following—which seems to call the objectivity of that activity into question. But let us set those issues aside.

In any case, there is a lot more to mathematics than simple calculations. Does cognitive command hold throughout the enterprize? The epistemic standard for serious assertion in professional mathematics is *proof*. So suppose that one mathematician, Pat, produces what she takes to be a proof of a mathematical proposition *S*; and that another mathematician, Chris, demurs from *S*, even after being presented with Pat's purported proof. The disagreement between Pat and Chris need not be over the conclusion of the purported proof. They differ over whether the purported proof is good—whether it establishes its conclusion. Pat believes *S*, on the basis of her purported proof, and Chris does not believe *S*, as he rejects the correctness of the purported proof. He may either think that *S* is false or else he may be agnostic about it.

In these circumstances, our question here is whether we can be sure, a priori, that at least one of our mathematicians exhibits a cognitive shortcoming—assuming that the disagreement is genuine, which is another matter on the table here.

In the world of professional mathematics, disputes like this happen. Two referees may disagree whether the argument in a submitted article does in fact prove its conclusion, with the competence of neither referee (nor the author of the paper) in doubt. But there is nothing special about mathematics here. There will be similar 'blameless' disagreements in any sufficiently complex area of discourse. To give cognitive command a chance of serving as an axis of *objectivity*, and of it helping to shed some light on the status of pure mathematics, we have to idealize on the cognizers.[2]

---

[2] I noted above that, according to Wright, cognitive command is an axis of objectivity only if the discourse is epistemically constrained—only if there are no unknowable

The idealizations in question here are familiar. We assume that our subjects have unlimited lifetimes, materials, memory and attention spans. These are the same idealizations invoked in the mathematical theory of computability, and in mathematical logic generally. Our discussion now threatens to turn back on itself. To assess whether cognitive command holds in mathematics, and thus whether mathematics is objective on this axis, we turn to some mathematics, to negotiate the idealizations. Well, I guess we all know that philosophy is a holistic enterprise.

For what it is worth, the pursuit of mathematics seems to me more like discovery than invention, more like getting at the truth than expressing an attitude. It also seems to me, as an interested outsider, that for the most part, the mathematical community shows a remarkable tendency toward convergence, perhaps more so than in any other area. At least at this point in history, disputes concerning the correctness of a given argumentation do not last forever. Unless the parties simply lose interest—surely a non-cognitive matter—actual disagreements seem to get resolved to everyone's satisfaction. Everyone comes to agree that a certain step is invalid, for example, or that a certain premise is suppressed, or that the argument is correct after all. This is surely evidence that cognitive command holds in the world of professional mathematics, at least when things are suitably idealized.

Define a *formal proof* to be a sequence of sentences, in a formal language, in which every step is either an explicitly noted axiom, premise or assumption, or else follows from previous steps via what, for the author at least, is a primitive rule of inference, one so basic that there is no sense in breaking it down further. In actual mathematics, it is not always clear whether a given proof has a unique formalization. Is there an objective fact of the matter, independent of judgements and the like, concerning how to formalize any given piece of published mathematics? Let me concede that an opponent of objectivity has some room to manoeuver here. The move from actual mathematical discourse—which is what we care about after all—to the fully formalized proofs of the idealized mathematicians may not be governed by fully objective standards. But, as noted above, full objectivity is not likely to be on the cards in any case. We are exploring the extent to which mathematics is objective, on this one axis.

Let us return to our imaginary mathematicians Pat and Chris, now suitably idealized. Assume that Pat puts forward a fully formalized proof Π of a mathematical proposition S and that Chris rejects the proof, demurring from its conclusion.

---

truths, and I asked us to assume that mathematics is epistemically constrained. To give that a chance, we also have to focus on idealized mathematicians. There are surely some true propositions that no flesh and blood mathematician can ever know, simply because the proposition requires a calculation that is too long to be completed before the Sun goes cold.

Pat and Chris presumably agree on what formula appears on each line of the purported proof Π. A disagreement there would involve a cognitive shortcoming on the part of one of them; he or she does not have good eyesight. So either Pat and Chris disagree over one of the axioms, assumptions or premises of Π, or they disagree over the validity of one of Pat's primitive rules of inference.

In mathematics, a difference over a premise or axiom is, prima facie, not really a disagreement. The two mathematicians are just talking past one another. Pat is working in a certain structure (or type of structure), character-ized in part by the premises of the derivation Π, while Chris prefers to work in a different structure. A mathematician who demurs from the Pythagorean theorem, because he does not assume the parallel postulate, is not in real disagreement with a Euclidean who does. The two of them work in different theories, with different subject matters.

Of course, mathematicians did not always think this way. Supposedly, they once saw the issue concerning geometry as concerning the structure of (physical) space or intuitions concerning perception, or something else along these lines. Alberto Coffa (1986, p. 8) describes the historical transition:

> *During the second half of the nineteenth century, through a process still awaiting explanation, the community of geometers reached the conclusion that all geometries were here to stay... [T]his had all the appearance of being the first time that a community of scientists had agreed to accept in a not-merely-provisory way all the members of a set of mutually inconsis-tent theories about a certain domain... It was now up to philosophers... to make epistemological sense of the mathematicians' attitude toward geome-try... The challenge was a difficult test for philosophers, a test which (sad to say) they all failed....*

I take it that on the present scene, if Pat and Chris differ only over premises or axioms, then they do not disagree at all. In effect, their words mean different things. The two of them simply work in different structures. This explains why mathematical theories are not discarded as false when they become unusable in science, at least not now. Michael Resnik (1997, p. 131) calls the phenomenon 'Euclidean rescue'.

Things may not be this neat if the disagreement concerns a more foun-dational matter. Suppose that the ultimate conclusion of the 'disputed' proof is a proposition of real analysis, but that Pat's proof invokes a set-theoretic principle, such as the continuum hypothesis or a large cardinal hypothesis, and Chris rejects that set-theoretic principle. This would naturally focus the dispute on the background set theory. One can invoke a Euclidean rescue there as well, and say that Pat and Chris work in different structures, just because their background set theories differ. Pat works in analysis-plus-set-theory-*A*, while Chris prefers analysis-plus-set-theory-*B*. This sort of resolution is not quite as comfortable as it was with, say, Euclidean and non-Euclidean geometries, due to the pervasiveness of set-theoretic notions throughout mathematics, and the foundational role of set theory (see Maddy, 2007, pp. 358–360). The matter can be murky, and it goes beyond the scope of this paper. We'd have to consider whether mathematics can have more than one foundation, and, if it does, how we'd study the relationships between the foundations.

In any case, a dispute concerning the set-theoretic background is much like our one remaining possibility, that the difference between our mathematicians' attitudes traces to their logic. Suppose that Chris demurs from a rule of inference that, for Pat, is so primitive that it cannot be broken down further. To focus on an example, let us suppose that Pat's proof invokes instances of excluded middle, and Chris rejects that, since he is an intuitionist. This raises the question of the objectivity of logic, which could (and did) take up another lengthy paper (Shapiro, 2000). The details go beyond present concern, but I suggest that with some qualifications, logic, too, passes the letter of all of Wright's tests for objectivity, with the possible exception of epistemic constraint. But it is not clear what one should conclude from that.

One potentially troubling matter here there is that most of Wright's axes of objectivity are formulated in terms of logic, and so it seems that one can deploy the various axes only *after* one has settled on a logic. That is, the various axes of objectivity *presuppose* a logic (although it is left open just which logic is presupposed). So it hard to see the extent to which we can even ask if logic is objective, using Wright's framework.

One might invoke Euclidean rescue with logic, taking an eclectic attitude toward it. The thesis would be that, say, classical analysis and intuitionistic analysis are two different subjects, and are no more in conflict with each other than Euclidean and non-Euclidean geometries. If we go down this route, then in a sense all of mathematics—once suitably idealized—reduces to calculation. It is just a matter of what conclusions can be drawn in various deductive systems. There is no content to mathematics beyond that. We save cognitive command, but at the cost of there not being any interesting, genuine disputes among mathematicians. The supposedly 'disputing' parties do not speak the same language. We seem to have manoeuvered ourselves into some sort of formalism concerning mathematics—at least once it is suitably idealized so that we can apply the axes of objectivity. Of course, we are still putting Wittgensteinian issues of rule-following aside.

Let us briefly explore the alternative position, that the classical mathematician (Chris) and the intuitionist (Pat) do have a genuine disagreement with each other. And then we can ask if the matter is objective, along the axis of cognitive command. More murky waters lie here.

Our question concerns whether one of our idealized disputants, Chris or Pat, exhibits a cognitive shortcoming. 'Inferential error' appears in the list of shortcomings that Wright gives in the above characterization of cognitive command.[3] Surely Pat *accuses* Chris of inferential error. Chris invokes excluded

---

[3] Recall: 'A discourse exhibits Cognitive Command if and only if it is a priori that differences of opinion arising within it can be satisfactorily explained only in terms of

middle, and that, for Pat, is an error. However, Wright does not seem to have this in mind. He takes 'inferential error' to be result of 'inattention or distraction', not a deep disagreement concerning the nature of logical consequence itself. It is part of the idealizations invoked here that neither Chris nor Pat has a failure of concentration. So if cognitive command is to hold here, we will have to find some other sort of cognitive shortcoming to attribute to one or the other of our idealized mathematicians.

Arguably, logic-choice is a holistic enterprise, although this is yet another controversial matter that cannot be addressed here fully. Resnik proposes an adaption of the programme of 'wide reflective equilibrium' formulated by Nelson Goodman and used by John Rawls for an account of justice:

> One starts with one's own intuitions concerning logical correctness (or logical necessity). These usually take the form of a set of test cases: arguments that one accepts or rejects, statements that one takes to be logically necessary, inconsistent, or equivalent... One then tries to build a logical theory whose pronouncements accord with one's initial considered judgements. It is unlikely that initial attempts will produce an exact fit between the theory and the "data"... Sometimes... one will yield one's logical intuitions to powerful or elegant systematic considerations. In short, "theory" will lead one to reject the "data". Moreover, in deciding what must give, not only should one consider the merits of the logical theory per se... but one should also consider how the theory and one's intuitions cohere with one's other beliefs and commitments, including philosophical ones. When the theory rejects no example one is determined to preserve and countenances none one is determined to reject, then the theory and its terminal set of considered judgements are in... wide reflective equilibrium.
>
> Resnik (1997, p. 159)

There is another troubling circularity here. In order to see if one is in reflective equilibrium, she must reason logically. She must draw logical consequences of her logical theory, to see if it coheres with her intuitions and other 'data'. There can be no rigorous characterization of reflective equilibrium that is neutral on the choice of logic. We can only speak of reflective equilibrium for a given logic.

Still, perhaps one of our disputants may fail to be in reflective equilibrium, by his or her own lights. That would surely count as a cognitive shortcoming on the part of that disputant. But can we be sure that this always happens? That is, can we *rule out* cognitively blameless disagreement?

"divergent input", that is, the disputants working on the basis of different information (and hence guilty of ignorance or error...), or "unsuitable conditions" (resulting in inattention or distraction and so in inferential error, or oversight of data, and so on), or "malfunction" (for example, prejudicial assessment of data ... or dogma, or failings in other categories ...' (Wright, 1992, p. 92)

It seems not, although it is hard to see how one might construct an argument for this. It seems to be possible that our two idealized mathematicians, Pat and Chris, are both in reflective equilibrium, each by the lights of his or her own logic. If so, how can we neutral observers accuse one of them of cognitive shortcoming?

There is something troubling about the whole issue concerning the objectivity of logic. Any serious dispute in any area of discourse is going to involve logic. All disputants, in all areas, are themselves reasoners, and come to their respective conclusions in part by drawing inferences. Given how pervasive logic is, disagreements or differences about logic are certain to result in disagreements or differences everywhere. If logic fails to be objective, can there be any objectivity anywhere?

I apologize for failing to come to a crisp conclusion concerning the objectivity of logic and, to that extent, the objectivity of mathematics. Given the central role of logic in our theorizing, it is hard to separate it out for sharp treatment. Any attempt to characterize how the question of objectivity is to be adjudicated will presuppose logic.

The situation can be made more palatable if we recall the Kant–Quine thesis, the idea that there is no way to sharply separate the parts of our best theories that are due to the way the world is and the parts that are due to the way that we, the human cognitive agents, are. One fallout, I submit, is that there is no sense to asking for *complete* objectivity. So some feature that compromises the objectivity of a given area of discourse does not eliminate objectivity altogether. This is the nature of the holistic beast.

Later in his book, Wright (1992, p. 144) adds some qualifications to the formulation of cognitive command. It looks like the qualifications are meant to deal with matters like holistic adjudication, in our struggle for reflective equilibrium. Wright says that a discourse exerts cognitive command if and only if

> it is a priori that differences of opinion formulated within the discourse, unless excusable as a result of vagueness in a disputed statement, or in the standards of acceptability, or variation in personal evidence thresholds, so to speak, will involve something which may properly be described as a cognitive shortcoming.

That is, Wright holds that blameless disagreement that turns on vagueness, standards of acceptability and the like, does not undermine cognitive command. Why are there these exceptions? What happened to the original motivation for the criterion of cognitive command, which did not mention vagueness, evidence thresholds, or the like, one way or the other? Is this an instance of what Imre Lakatos (1976) calls 'monster-barring'? We find some parts of our theory that do not seem to fit, and so we just exclude them.

Although Wright does not put it this way, I suggest that the exceptions listed in the nuanced version of cognitive command are in line with the Kant–Quine theme. On Wright's views of vagueness—and mine

(Shapiro, 2007a)—vague terms are response- or judgement-dependent, at least in their borderline regions. He writes that it is 'tempting to say...that a statement's possessing (one kind of) vagueness just *consists* in the fact that, under certain circumstances, cognitively lucid, fully informed and properly functioning subjects may faultlessly differ about it.' Yet, robustly objective areas of inquiry, such as natural science, are conducted with vague terms. That, alone, cannot undermine their objectivity, unless one takes an all-or-nothing approach to that question. Similarly, it is plausible that 'standards of acceptability' and especially 'variation in personal evidence thresholds' lie closer to the 'human' side and further from the 'world' side of the mix. Surely, a conservative scientist, one who is more cautious in putting forward claims, need not have any cognitive shortcoming with respect to a slightly more speculative colleague, nor vice versa. So a disagreement that is traced to that difference need not undermine cognitive command.

The same may go for the more foundational matters concerning mathematics and its logic. Disagreements that turn on holistic considerations may end up being adjudicated, in part, by matters of taste concerning, for example, what a given theorist finds to be elegant or simple. One mathematician may prefer the subtle distinctions and sharper bounds produced by constructive mathematics while another might go for the unity and, in some way, simplicity and tractable meta-theory of classical mathematics. That is, it just may be that certain foundational matters are negotiated closer to the 'human' than the 'world' side of the web. Even if this is so, it does not follow that mathematics is not objective, even a paradigm of objectivity, one of the standards by which we measure other discourses.

Comment on Stewart Shapiro's 'Mathematics and objectivity'

Gideon Rosen

———=⊃◦◦◦⊂———

Stewart Shapiro approaches the question of the objectivity of mathematics indirectly. Instead of asking whether the mathematical facts somehow *depend on human thinking*, Shapiro asks whether mathematical discourse exhibits what Wright calls *cognitive command*. His chapter presupposes that these ideas are connected in a simple way:

> If a region of discourse fails to exhibit cognitive command, then the facts with which that discourse is concerned are not objective facts.

In this note I raise a doubt about this principle.

Suppose that you and I disagree about the merits of Wagner's *Ring*. You say it's a masterpiece; I say it's a bore. Suppose we both know the operas and the relevant musicological background very well, that neither of us is drunk or distracted, that our judgements are stable upon reflection, etc. Given all of this, we have a disagreement that may persist even though both parties have a firm grip on the underlying facts and neither has made a mistake in reasoning. Pressed to explain the disagreement, we may be reduced to saying that we disagree simply because we bring incompatible but equally coherent aesthetic sensibilities to the question.

An area exhibits cognitive command, roughly speaking, when every disagreement is the result either of *divergent input*—i.e., differences in the information available to the disputants—or some sort of cognitive malfunction or *mistake in reasoning*. The example suggests that aesthetic discourse fails to exhibit cognitive command and hence, given the above presupposition, that aesthetic facts are not objective facts. Shapiro argues (with some qualifications) that mathematical discourse passes the test with flying colours, and hence that so far as this criterion is concerned, there is no reason to doubt the objectivity of mathematics.

Is this right? Modern mathematics is ultimately a matter of proving theorems from axioms by means of rules. Any mathematical disagreement that is not due to a simple mistake is therefore traceable either to disagreement about the axioms or to disagreement about the rules of logic. Let us set disagreement about logic to one side. This is an important issue, but as Shapiro notes, it is hard to discuss within Wright's framework. If we focus on disagreement about the axioms, then the first thing to emphasize (as Shapiro does) is that while disagreement of this sort has occasionally surfaced in mathematics, e.g., the debate over the parallel postulate in geometry, modern mathematics has a standard way of finessing it. The idea is to treat the disputed axioms as clauses in the definitions of special mathematical structures. Where geometers might once have disagreed about the absolute truth or falsity of the parallel postulate, a modern mathematician will say: 'Some spaces are Euclidean, others aren't. The axiom holds in every Euclidean space, since conformity to the axiom is part of what makes a space Euclidean. But it fails in other spaces, examples of which are easily given'. On this account, it makes no sense to ask whether the axiom is true or false simpliciter, and so it makes no sense to disagree about its truth.

Apparent conflicts in mathematics can usually be dissolved in this way—but not always. One central concern of mathematics is to establish the existence of mathematical objects satisfying various conditions. Now an existence proof always requires at least one existence axiom, and if one examines real mathematics one finds that it is always legitimate simply to *assert* the existence of the natural numbers and of certain sets of natural numbers. These existential claims are not mere conditions or assumptions. When a mathematician proves the consistency of hyperbolic geometry by producing a model of the axioms in (say) $\mathbb{R}^3$, the content of his theorem is not '*If there are numbers, etc*, then a model of the axioms exists.' His proof establishes the existence of a model outright, and this means that it must involve the assertion of at least one existential axiom.

There is of course no real disagreement within mathematics about the existence of the natural numbers and certain sets constructed from them. But in applying the cognitive command test we are not constrained to focus only on real disagreement. Even if there were no actual disagreement about the merits of the *Ring* (thanks to uniform musical education), the mere possibility of the disagreement described above would suffice to show that aesthetic discourse flunks the cognitive command test for objectivity. By the same token, the mere possibility of disagreement about the existential claims of standard mathematics would suffice to establish the non-objectivity of at least part of mathematics, provided that disagreement was not traceable to 'divergent input' or 'mistakes in reasoning'.

Such disagreement is clearly possible. Mathematicians who have been acculturated in the normal way find the existential claims of basic arithmetic perfectly obvious and hence acceptable without proof. But we know that it is possible for a person to find these axioms anything but obvious. After all, some *philosophers* explicitly reject them on the grounds that (a) they have no intrinsic plausibility, and (b) every positive argument in their favour is unconvincing (Field, 1980; Leng, this volume). These philosophers typically point out, for example, that numbers would be invisible, non-physical entities of some sort, and that it is hardly evident *to them* that such things exist.

I want to imagine this as a genuine clash of sensibilities. Some people find the existential axioms obvious and so affirm them; others find them totally non-obvious and withhold assent. Is this a matter of 'divergent input'? Of course the axioms *strike them* differently. But if we treat this as a matter of divergent input in applying the cognitive command criterion, we shall have to say that our disagreement about Wagner is also a matter of divergent input, and that would trivialize the criterion. Is the disagreement due to a mistake in reasoning, or to some other cognitive malfunction? Perhaps; but if it comes down to a disagreement about the intrinsic plausibility of an axiom, it is hard to see why this should be so.

The possibility of a disagreement about the existential assumptions of standard mathematics suggests ordinary mathematics may not exhibit cognitive command. Should we conclude that mathematics is not objective after all? No. We should conclude instead that cognitive command is a flawed criterion of objectivity. Objectivity in the relevant sense is a metaphysical concept. To call a fact objective is to say that it does not depend in any interesting way on thought or language (Rosen, 1994). From the fact that a disagreement in mathematics might be traceable to variation in our sense of 'mathematical plausibility', *nothing follows* about the metaphysical status of its subject matter. If we discovered that disagreement about the existence of God were sometimes traceable to a difference in theological 'sensibility', would we conclude that God's existence (or non-existence) was somehow mind-dependent? We might conclude that *judgements* about God's existence are not strictly forced upon us by the evidence. But this is a point about the epistemic status of these judgements; it is not a point about the metaphysical status of the facts with which they are concerned.

Reply to Gideon Rosen

Stewart Shapiro

I tend to agree with the main thrust of Gideon Rosen's comment on my 'Mathematics and objectivity'. In particular, I agree that Crispin Wright's notion of cognitive command does not, by itself, provide a necessary and sufficient condition for objectivity. The purpose of my paper (and two others) is to test mathematics and Wright's various axes of objectivity against each other. However, I do think that there is something right about cognitive command as at least a defeasible criterion; it is a matter of articulating its scope and limits. I hold, with Wright, that objectivity is not a univocal notion. There are various aspects of objectivity, not all of which line up with each other.

In another paper, 'Objectivity, explanation, and cognitive shortfall' (forthcoming in a festschrift for Crispin Wright), I provide a thought experiment involving two scientists who disagree with each other, but are each in reflective equilibrium concerning the overall balance of evidence. It would be hard to fault either of them. Yet one would be loath to conclude that science generally is not objective. A referee for that paper pointed out that Wright's criterion, by itself, does not distinguish failures of cognitive command due to the non-objective nature of the subject matter (e.g., Rosen's example about Wagner's *Ring*) and failures due to the scantness of evidence, especially in such areas where evidence is evaluated holistically. Perhaps Rosen's examples concerning mathematical existence, and the general philosophical themes of fictionalism, can be understood similarly. The referee located this in the history of other failed attempts to demarcate cognitive significance.

Toward the end of his commentary, Rosen suggests that cognitive command looks in the wrong place: objectivity is a purely metaphysical matter, while cognitive command is a broadly epistemic criterion. I do not agree with that perspective, but this is not the place to pursue that general issue.

9

The reality of mathematical objects

Gideon Rosen

If the truth be known, there are no such things as numbers; which is
not to say that there are not at least two prime numbers between 15
and 20.

Paul Benacerraf, 'What numbers could not be' (1965)

The problem

The closing sentence of Paul Benacerraf's famous paper is a kōan: a bit of
seeming nonsense that points—or seems to point—to a deep truth. It is a
theorem of basic arithmetic that there are two prime numbers between 15 and
20. Anyone who accepts basic arithmetic must therefore agree that there are
two prime numbers between 15 and 20, and hence there are at least two numbers, and hence that *there are numbers*. And yet the idea that numbers are real
things—that the real world contains mathematical objects in the same sense in
which it contains guns and rabbits—can sound preposterous or confused. And
so we find philosophers straining to articulate a position of the following sort:

> *Of course there are numbers (and functions and sets and mathematical
> objects of other sorts). That's just mathematics, and we have no quarrel with
> mathematics. But in another sense—the* metaphysical *sense—there are no
> numbers. Numbers are not* real. *Numbers are not* Things.[1]

It should be obvious that such remarks are seriously puzzling as they stand.
Suppose your exterminator tells you that you have squirrels in your attic,
but then goes on to add that in the strict and philosophical sense *there are
no squirrels*. Or suppose an astrophysicist reports that there are three black

---

[1] For a recent statement of this idea, see Dorr (2008).

holes at the center of the galaxy, and then goes on to say, 'And by the way, black holes are not *real*; they are not *Things.*' If you are anything like me, you will find these remarks incomprehensible. And yet in the philosophy of mathematics there is a widespread sense that such formulations, while perhaps less transparent than one might like, must nonetheless make some sort of sense.

Let us call the thesis we seek to clarify *qualified realism* about mathematics. The view is a form of realism because it holds that there is a sense in which mathematical objects exist. It is a *qualified* realism because it holds that these mathematical objects are somehow metaphysically 'second rate'. The difficulty is to say what this qualification comes to.

I used to think that this could not be done in such a way as to yield a version of qualified realism worth discussing (Rosen and Burgess, 2005; Rosen, 2006). I now suspect that this pessimism was premature. My aim in this note is to present an account of what the qualified realist might have in mind—an account according to which the objects of mathematics compare unfavourably in certain metaphysical respects with certain paradigmatic Things: the objects of everyday experience, perhaps, or the objects of the physical sciences. I do not endorse the version of qualified realism that I shall discuss, but I believe it merits our attention. The aim here is simply to put the view on the table.

The case for minimal realism about mathematical objects

Before we begin, it may help to review the case for realism itself (Burgess, 1983; Burgess and Rosen, 1997; Rosen and Burgess, 2005).

By *minimal realism* about mathematical objects, I mean the claim that mathematical objects exist. I use the phrase 'mathematical object' as a catch-all to cover numbers, functions, sets, groups, spaces, models, vectors, categories, systems of equations, formal languages, and the other manifold items with which mathematics is distinctively concerned. To say that mathematical objects exist is just to say that there is at least one item of this sort, or equivalently, that one such thing exists.

Must I say what I mean by the word 'exist' in this context? I don't think so. The existential idioms—the predicate 'exists', along with quantificational expressions like 'there are...' 'there exist...', 'at least one...', etc.—are part of the everyday language of mathematics. These idioms are all equivalent in that language, and they are not ambiguous. That is why they can all be represented by a single symbol, $\exists$, in standard formalizations of mathematics. If you are reading this paper, you already understand that language, and I propose to rely on that understanding. So when I say that minimal realism is the thesis that mathematical objects exist, I might add that I mean to use the word 'exist' in its ordinary mathematical sense, the sense a high-school student or a professional mathematician has in mind when he says that there exist two solutions to some equation. This is not a definition, but it should suffice.[2]

So understood, minimal realism is not an esoteric metaphysical claim. It is a mathematical claim—a trivial consequence of the most elementary parts of mathematics. As noted, it is a theorem of arithmetic that there are two prime numbers between 15 and 20. This theorem entails that there are at least two numbers, and hence two mathematical objects. Anyone who accepts basic arithmetic therefore has no choice but to accept minimal realism.

Should we accept basic arithmetic? Are we justified in believing that there are two prime numbers between 15 and 20? In my view, the claims of elementary mathematics have the same status as the claims of common sense in other areas—e.g., the claim that we inhabit a world of real things that exist even when we are not aware of them, or the claim that other human beings have conscious mental lives rather like our own. Common sense is fallible, of course. But if a philosopher (or a scientist or anyone else) wishes to call the claims of common sense into question, he must give reasons for his doubts. And when it comes to basic arithmetic, such reasons simply do not exist. Arithmetic is obviously unassailable on mathematical grounds.[3] And if we treat ordinary scientific standards for the acceptance and rejection of theories as authoritative, the fact that every developed science takes mathematics utterly for granted is enough to show that the scientific enterprise broadly conceived has never thrown up doubts about arithmetic. If there are any grounds for doubt in this area, they must therefore be distinctively *philosophical* grounds.

I will not try to survey the arguments that philosophers have developed in this context, but I will say (very briefly) why I find them unpersuasive. The existing arguments are of two kinds. Some philosophers say that we should reject standard mathematics because numbers and the like would be queer things if they existed. And this is certainly true. If there is such a thing as the number 26, it is very different from the objects of everyday experience (tables, etc.) and the less-familiar objects disclosed by physics (quarks, etc.). But so what? To advocate the rejection of arithmetic on this ground is repose more confidence in a grand metaphysical scheme (sometimes called *physicalism*) according to which *absolutely everything* is rather like a table or a quark,

---

[2] I might simply have said that I use 'exist' and the other existential idioms to mean what they mean in ordinary English. As Quine stressed, there is no good reason to believe that these words mean one thing in mathematics and something else in other areas (Quine, 1960, §27). Numbers are plainly very different from tables and quarks and mental images. But if all of these things exist—i.e., if there are such things—there is no reason to suppose that they exist in different *senses*.

[3] Some philosophically minded mathematicians have expressed doubts about certain parts of classical arithmetic—the impredicative or non-constructive parts (for example, Nelson, 1986). These specialized disputes need not concern us. Constructive arithmetic involves existence claims, e.g., the claim that there are two prime numbers between 15 and 20.

than in the arithmetical claim that there are two prime numbers between 15 and 20. But if it is a matter of choosing between grand metaphysical schemes and basic arithmetic, it is clear to me that it is the metaphysics that ought to budge.

Other philosophers reject basic arithmetic because they think that if numbers and other mathematical objects existed, there would be no way for us to know anything about them. This sort of claim is typically supported by a general philosophical theory of knowledge, according to which knowledge requires some sort of interaction between the would-be knower and the object of his inquiry (Benacerraf, 1973). These theories were originally developed to account for empirical knowledge, and they may be useful for that purpose. But if the philosopher insists that they hold in complete generality, he faces a glaring difficulty. These restrictive theories typically entail that the usual ways of fixing opinion in mathematics—calculation, proof, informal mathematical argument—could not possibly be sources of knowledge (since they do not involve causal intercourse with the numbers), and hence that someone who has come to believe in the ordinary way that $235 + 657 = 892$ does not really *know* that $235 + 657 = 892$. But then the question arises: Why isn't this simply a *counter-example* to the philosopher's theory? In other areas, when a philosophical theory is incompatible with an otherwise uncontroversial fact, the normal response is to reconsider the theory or to restrict its scope. A philosophical theory of knowledge must accommodate the manifest fact of mathematical knowledge. If the theory clashes with this fact, so much the worse for the theory.

This is just a rough sketch of an intricate dialectic, but the main tactic should be clear. The central claims of mathematics are unassailable by all pertinent mathematical, scientific and commonsensical standards. Any philosophical challenge to these claims is thus a sceptical challenge of a certain special sort, one that depends on bringing distinctively philosophical principles to bear on questions that are not themselves purely philosophical. Sceptical challenges of this sort are notoriously weak. When speculative philosophy contradicts settled science or common sense, the normal response—and, I believe, the reasonable response—is to suspect that it is the philosopher who is mistaken. This is not a firm principle, but it is a good rule of thumb. And if we follow it in this case we have no choice but to allow that since there are two prime numbers between 15 and 20, mathematical objects therefore exist.

Qualified realism: an example

Minimal realism, then, is not just a philosophical claim; it is a (trivial) bit of settled mathematics. And yet for many philosophers the view conjures up a repugnant picture according to which mathematics is like zoology: a science whose aim is to describe the curious behaviour of a special class of things, the main difference being that in the mathematical case the objects of interest are infinite in number and totally invisible.[4] It is this thought, among others, that has led philosophers to grope in the direction of the view we have called *qualified* realism: the view that numbers are not really *things* in the sense in which lions and tigers are *things*. Our task is to give a sense to this dark thought.

We begin with some examples of views that seem to encourage this sort of verbiage. The examples are all versions of *reductionism* about arithmetic. In general, a reductionist thesis in this area holds that the arithmetical facts are somehow *grounded in* or *amount to nothing over and above* facts of another, more fundamental, kind. (These italicized phrases raise an eyebrow, and we shall return to their interpretation.) Now some reductionist proposals of this sort have no interesting metaphysical implications. If a theorist identifies the individual numbers with pure sets—say, by claiming that 0 is the empty set, and that the successor of a number *n* is the set whose sole member is *n*—then every fact of arithmetic may be reduced to a fact of set theory. And yet this sort of reduction by itself has no tendency to impugn the Reality of the numbers. After all, it is consistent with this view that *sets* are as robust and thing-like anything could be, and since the view identifies numbers with sets, it is therefore consistent with unqualified, full-strength realism about the numbers.

We get a more interesting form of reductionism when we consider proposals that purport to reduce the *truths* of some part of mathematics to truths in some more basic idiom without identifying the *objects* of mathematics with objects recognized in the more basic theory. As an illustration, consider *formalism* in the philosophy of arithmetic. The formalist's central thought is that arithmetic is not ultimately concerned with an extralinguistic domain of things. Rather, insofar as arithmetic has a proper subject matter, it is the *language* of arithmetic itself and certain formal relations among its sentences.[5] Here is a simple version of the view. Let PA be the usual formalization of elementary arithmetic: first-order Peano arithmetic.[6] Let PA_ω be this theory supplemented

---

[4] 'Arithmetic as the natural history (mineralogy) of numbers. But *who* talks like this about it? Our whole thinking is penetrated with this idea.' (Wittgenstein, 1956: 1967, IV, p. 11)

[5] The view has few contemporary adherents among philosophers, though mathematicians often find it congenial. See Curry (1951) for one version. For an exposition and assessment of Frege's celebrated objections to formalism, see Resnik (1980).

[6] The axioms include the basic principles governing the successor function:

0 is a number
0 is not the successor of any number
Every number has a successor
No two numbers have the same successor;

the recursion equations for addition and multiplication;

For any number x, $x + 0 = x$
For any numbers x and y, $x +$ the successor of $y =$ the successor of $(x + y)$

with an infinitary rule of inference—an omega rule—that permits the inference from an infinite list of premises $A(0)$, $A(1)$, ... $A(n)$ ... to the universally quantified conclusion: *For all numbers x, A(x)*. PA_ω is obviously a sound theory. The axioms are true, and the rules of inference preserve truth. More importantly, PA_ω is also a *complete* theory, in the sense that every sentence A in the language of arithmetic is such that either A or its negation is provable in PA_ω.[7] This means that anyone who accepts standard arithmetic should accept the following equivalence:

For any sentence A in the language of arithmetic, A is true if and only if A is provable in PA_ω.

Taken by itself, this equivalence is a mathematical fact with no special metaphysical significance. Even the most unreconstructed Platonist should accept it. But now consider the formalist's characteristic claim:

For any true arithmetical sentence A:
A is true *in virtue of* the fact that A is provable in PA_ω; or
What *makes* A true is the fact that it is provable in PA_ω; or
A's truth is *grounded in* the fact that A is provable in PA_ω;
A's truth *consists in* the fact that A is provable in PA_ω.

The italicized idioms here are not part of the official vocabulary of mathematics. Mathematics assures us that claims about the natural numbers are *equivalent* to claims about the formal provability of certain sentences in a certain formal system. But the formalist's distinctive claim is that the arithmetical facts are somehow grounded in these proof-theoretic facts, and that the proof-theoretic facts are therefore, in a corresponding sense, more fundamental. From the standpoint of mathematics, this claim is extracurricular. It is a distinctively philosophical claim about the metaphysics of arithmetic.

Note that even though the reductive formalist regards certain linguistic items (sentences and formal systems) as more fundamental than the numbers, he cannot deny that numbers exist. Since the existence theorems of arithmetic (e.g., *there are two primes between 15 and 20*) are all provable in PA_ω, the

---

For any number x, $x \times 0 = 0$
For any numbers x and y, $x \times$ the successor of $y = (x \times y) + x$;

and an axiom scheme for mathematical induction:

If 0 is F, and if for all x, F(x) implies F(the successor of x), then every number is F.

[7] The completeness of PA_ω is consistent with Gödel's famous incompleteness theorem. Gödel's theorem shows that any consistent theory that includes basic arithmetic must be incomplete, *provided the theory is recursively enumerable*. Roughly speaking, a theory is recursively enumerable when there exists a finite mechanical procedure—a computer programme—for listing its theorems. PA_ω is not recursively enumerable in this sense, so Gödel's theorem does not apply.

formalist must concede that these theorems are true. But as we have seen, if this particular theorem is true, two prime numbers exist. So the formalist must say (as emphatically as one likes) *THERE ARE NUMBERS; NUMBERS EXIST.* What matters for our purposes is what he goes on to say, viz., that *all it takes* for numbers to exist is for certain *sentences* to be derivable from certain others in a certain formal calculus. It is at this point that he begins to sound like something less than a full-strength realist about the numbers.

One is tempted to explain this as follows. If the numbers were first-class things then our claims about them would be true, when they are true, in part because the numbers are as they are. That's how it is in zoology. Penguins are real things, and true claims about them are rendered true, at least in part, by the birds themselves. In arithmetic as the formalist understands it, by contrast, a statement about the numbers is true (when it is) *because it is derivable from certain axioms in accordance with certain formal rules.* The numbers themselves play no role in grounding the truths of arithmetic. In fact it is natural to suppose that it is the other way around. If we ask the formalist to tell us why Euclid's theorem on the infinity of the primes is true, he will say that it is true because it is a theorem of PA_ω. And if we then ask him why there are infinitely many prime numbers—that is, if we ask him a question about the numbers themselves and not about the truth of a certain sentence— he may say: there are infinitely many prime numbers because the sentence '*There are infinitely prime numbers*'[8] is true. On this account (which goes a bit beyond the views explicitly attributed to the formalist above)[9], not only do the mathematical objects play no role in making our mathematical theories true, the objects themselves exist only because our claims about them are true for other reasons. And as soon as one hears this one has the palpable sense that formalism amounts to a form of *qualified* realism about the numbers.

Another example: modal structuralism

As another example, consider a version of structuralism in the foundations of arithmetic. The structuralist begins from an important insight, namely, that whenever we prove a theorem about the natural numbers, we in fact prove a more general theorem about any collection of objects isomorphic to the numbers. The natural numbers in the standard order

$$0, 1, 2, \ldots$$

---

[8] ... or its surrogate in the language of formal arithmetic.

[9] To be explicit, we are now imagining a version of formalism that involves two schematic claims:

For true arithmetical statements A:

(i) The fact that 'A' is true is grounded in the fact that 'A' is a theorem of PA_ω;

(ii) The fact that A is grounded in the fact that 'A' is true.

exemplify a distinctive pattern: a discrete linear order with no last element, in which every item has only finitely many predecessors. If the succession of Roman emperors had gone on forever, the sequence

Augustus, Tiberias, Gaius,...

would have been another instance of this pattern. We call any ordered collection of this sort an *omega sequence*. In general, an ordered collection is a pair $(X, \prec)$ consisting of a set X and a relation $\prec$ on that set. In the hypothetical example, the Roman emperors together with the temporal relation *x's reign precedes y's reign* constitutes an omega sequence, as do the natural numbers together with the relation *x is less than y*.

In order to state the relevant fact about arithmetic, we need one more bit of received wisdom. It is well known that given a suitable logical background, every arithmetical claim that can be formulated in the elaborate technical language of modern number theory can be expressed in a stripped-down language in which the only primitive symbols are N (for *natural number*) and $<$ (for the standard *less than* relation). If A is an ordinary arithmetical sentence (e.g., *there are two prime numbers between 15 and 20*), let's call its translation into this stripped-down language $A[N, <]$.

The pertinent theorem is then as follows:

For any claim A in the language of arithmetic, $A[N, <]$ if and only if, for any omega sequence $(X, \prec)$, $A[X, \prec]$.

Here $A[X, \prec]$ is the result of replacing the specifically arithmetical vocabulary in $A[N, <]$—that is, every occurrence of N and $<$—with *variables* X and $\prec$ ranging over sets and relations on those sets. Thus if $A[N, <]$ is a sentence in the language of arithmetic that says the natural numbers exhibit a certain arithmetical feature, the right-hand side of this equivalence says that *any* omega sequence—even one whose elements are Roman emperors—exhibits a corresponding purely structural feature. The theorem thus entails that every arithmetical claim about the numbers is equivalent to a perfectly general claim about omega sequences.

Now at this point we must note an assumption of this little theorem. An omega sequence is an infinite set, so if there are no infinite sets there are no omega sequences. But if there are no omega sequences, then *every* instance of the right-hand side of the equivalence is trivially true. This means that the equivalence holds only if infinite sets actually exist, and one may wish to avoid this assumption. One way to do this is to consider a somewhat different equivalence. Even if there are in fact no infinite collections, there *could have been*. The Roman Empire did not endure forever, but it *could have*. With this in mind, we might consider the following equivalence.

$A[N, <]$ if and only if *as a matter of necessity*, for any omega sequence $(X, \prec)$ $A[X, \prec]$.

Even if there are no infinite sets in the actual world, the claim that any *possible* omega sequence has a certain structural feature $A(X, \prec)$ is non-trivial (on the assumption that omega sequences *are* possible). The theorem affirms that this modal claim is equivalent to the ordinary mathematical claim A with which we began. (A modal claim is a claim about what is possible or necessary.)

Now so far this is just a bit of uncontroversial (if somewhat unfamiliar) mathematics. Anyone who accepts standard arithmetic should accept this equivalence. The structuralist's characteristic philosophical claim is that the truths of arithmetic *reduce to* or *are grounded in* general claims about all possible omega systems.[10] Return to our example, the claim that there are two prime numbers between 15 and 20. The structuralist accepts this claim, since he accepts ordinary arithmetic as it stands. And this means that he (like the formalist) cannot deny that numbers exist. His distinctive claim is that this fact about the numbers is grounded in the fact that if there were an omega sequence of whatever kind, it would have a certain complex structural property—very roughly, the property of having two 'prime elements' in between in its 16th and 21st elements.[11] And here is the crucial point: on the face of it, this latter fact is not a fact about a distinctive kind of object. It is not a fact specifically about numbers, nor is it a fact specifically about Roman emperors. Indeed there is a sense in which it is not a fact *about* anything at all. After all, a conditional modal claim of the form *if there were an infinite set of a certain kind, it would exhibit such and such features* does not affirm the existence of anything in the actual world. According to one sort of structuralist, the truths of arithmetic are made true by modal facts of just this sort. And it is not hard to see why someone who takes this view might be tempted to say that there is a sense in which the numbers are not real *things*. The truths about penguins are made true by the birds and their behaviour. The truths about the numbers, by contrast, are not really made true by the numbers; rather they are made true by general conditional facts in which the numbers themselves make no appearance.

A framework for formulating the proposal

Examples could be multiplied, but the pattern should be clear. Mathematical objects of a given sort (e.g., the natural numbers) are said to be *reducible* when every truth about them obtains in virtue of some truth in a more fundamental idiom in which the objects in question do not figure. In our examples, the reducing facts—the facts about provability in PA_ω, or the facts about every possible omega sequence—involve no explicit reference to (or quantification

---

[10] This is a version of 'eliminative' structuralism modelled on Benacerraf (1965), Putnam (1967), and Hellman (1989). A different view, also called structuralism, is defended by Resnik (1997) and Shapiro (1997).

[11] This assumes that the natural numbers begin with 0.

over) the numbers, and yet every fact about the numbers is said to obtain in virtue of some such fact. If a view of this form is correct, then while the numbers are perfectly real in one sense—there are such things—they may be 'unreal' in the sense that when we inspect the facts that ultimately ground the truths of arithmetic, we find no numbers of any sort. This suggests a slogan: *real things do not disappear upon reduction.* One way to reject the 'mathematical objects picture', as Hilary Putnam calls it (Putnam, 1967), is to hold that mathematical objects are unreal in just this sense.

The discussion thus far trades one mystery for another. We began by wondering what it could mean to say that numbers and other mathematical objects aren't really *things*. The proposal is to explain this notion in terms of another: the idea that truths of one sort *reduce* to truths of another sort. And yet this notion is notoriously problematic. Certainly there has been a great deal of confusion about what it means to say that truths about (say) our conscious mental lives reduce to truths about (say) the physiological processes in our brains and bodies. There are many conceptions of reduction, and hence many ways to interpret our proposal. Rather than survey the options, I will simply sketch one conception that strikes me as especially useful for our purposes.

As I understand it, reduction is a relation among *facts*—not among sentences or statements, but rather among the facts or states of affairs those sentences purport to describe. Reduction is thus a metaphysical relation, not a semantic one. To say that facts of one sort reduce to facts of another sort is not to make a claim about the meanings of words; it is to make a claim about the facts themselves, which typically obtain quite independently of our capacity to speak and think about them.

For present purposes we should think of facts as complex entities built up from objects, properties, relations and various other items in much the same sense in which sentences are built from words. For example, the fact that $2 + 3 = 5$ is a complex that might be represented as follows:

$$[= (+(2, 3), 5)]$$

Here the numbers 2 and 3, along with the identity relation and the operation of addition are literally *constituents* of the fact, just as the words '2' and '3' are constituents of the sentence '$2 + 3 = 5$'.[12] (In what follows, a sentence in square brackets names a fact whose structure corresponds to that of the enclosed sentence.)

My main substantive assumption is that facts stand in a basic relation of *grounding*. There is no standard English word for this relation, and there is no settled bit of philosophical terminology that picks it out. But we have a number of familiar idioms that point in the right direction, as when we say that one fact

---

[12] For my purposes, facts might be identified with true structured propositions of the sort originally described by Bertrand Russell (1905).

obtains *in virtue of* another, or that one fact *makes* another fact obtain. Some examples may help:

Disjunctive facts are grounded in their true disjuncts. It is a fact that I am now either in New Jersey or in Cambridge. The fact obtains (as it happens) in virtue of the fact that I am in New Jersey. If I had been in Cambridge, the same fact would have obtained for a different reason.

Existential facts obtain in virtue of their instances. It is a fact that *someone* spilled the milk. As it happens, this fact is grounded in the fact that *Fred* spilled the milk. If someone else had spilled it, the same fact would have obtained for a different reason.

Facts about the *determinable* features of things are grounded in more *determinate* facts. A certain ball is red in virtue of being (say) crimson; the particle has a mass of between 10 and 20 MeV in virtue of the fact that its mass is exactly 17.656 MeV.

Facts involving *definable* properties and relations are grounded in their 'definitional expansions'. To be a square is, by definition, to be an equilateral rectangle (or so we may pretend). Given this, if ABCD is a square, then it is square in virtue of the fact that it is equilateral and rectangular. That is what *makes* the figure in question a square.

Supervenient facts are typically grounded in the facts upon which they supervene, even if we cannot state the patterns of dependence in a systematic way. The fact that the US trade deficit with China in 2008 was roughly $117 billion supervenes upon a vast mosaic of particular facts about individual economic transactions, and perhaps ultimately on some vaster array of facts about the quarks and electrons that composed the people who did the buying and selling. Macroeconomic facts obtain in virtue of these lower-level facts, even if it is impossible in practice, and perhaps even in principle, for us to spell out the micro facts in virtue of which any given macro fact obtains.

These examples bring out two important features of the grounding relation. First, it is a form of necessitation. The facts that ground a given fact *entail* the fact they ground as a matter of absolute necessity. This distinguishes the grounding relation from certain forms of causal or nomological determination. There is no doubt a sense in which effects are grounded in their causes: the cause *makes* the effect occur, and so on. But as Hume noted, it is always possible for the cause to occur without the effect. The grounding relation that interests us involves a much more intimate form of dependence, as the examples show.

Second, the grounding relation is an explanatory relation. To cite the facts that ground a given fact is to give information about *why* that fact obtains. To suppose that the grounding relation is an objective relation, as I do, is therefore to suppose that there are objective facts about the explanatory order (which is not to say that the practice of *giving* explanations is always a matter simply of reporting these objective facts).

These remarks do not amount to a definition of the grounding relation. My own view is that the relation is too basic to admit of definition. More might be said by way of informal explication, of course, but I hope that this much will suffice for our present purposes.[13]

The proposal

A reductionist proposal in the philosophy of mathematics is a claim to the effect that every mathematical fact of a certain kind—e.g., every truth of arithmetic, or every truth of set theory—is ultimately grounded in facts of a rather different kind, e.g., acts about formal provability, or about what would have been the case if there had been an infinite sequence of objects.[14] We have considered two examples in which a reductionist proposal of this sort seems to imply that the objects of arithmetic are somehow less real or less 'thing-like' than certain other things. And in both of these proposals the reduction has had a distinctive character: the objects of the higher-level theory do not figure in the more fundamental facts to which that theory has been reduced. This suggests a natural strategy for explaining the metaphysical thesis of qualified realism.

Let us say that a fact is *fundamental* if it is not grounded in further facts, and that a *thing* is fundamental if it is a constituent of a fundamental fact. Then we might identify full-strength realism about mathematics with the thesis that some mathematical objects are fundamental things. This is certainly the view of the hardcore Platonist for whom the numbers are *sui generis* abstract substances—invisible luminous spheres arrayed in Plato's heaven. But it is also the view of more moderate Platonists who regard arithmetic, or perhaps some more comprehensive theory like Zermelo-Frankel set theory, as an autonomous body of truths not grounded in anything more basic. The forms of qualified

---

[13] The material in this section is developed in Rosen (2010).

[14] It is worth noting that the reductionist may also make a stronger claim, viz., that for every mathematical proposition p, for p to be the case *just is* for q to be the case—where q is a proposition involving objects of a more fundamental sort. Consider a standard scientific reduction: for it to be the case that x is hotter than y *just is* for it to be the case that the mean kinetic energy of the particles composing x is greater than the mean kinetic energy of the particles composing y. We may say that the fact about temperature *strongly reduces* to the fact about mean kinetic energy. Strong reductions of this sort always entail claims of necessary equivalence: if p strongly reduces to q, then as a matter of necessity, p is true if and only if q is true. That is enough to show that strong reduction differs from the grounding relation that we have been discussing. $[p$ or $q]$ may be grounded in $[q]$, but no one would say that for p or q to be the case *just is* for q to be the case. In general, $[p]$ may be grounded in $[q]$ even though $[p]$ does not strongly reduce to $[q]$; but not vice versa. If $[p]$ strongly reduces to $[q]$, then $[p]$ is grounded in $[q]$. In Rosen (2009) this claim is called the *Grounding-Reduction Link*.

realism that we have discussed are all opposed to mathematical realism of both these sorts.

Stated more generally, the proposal for explaining the puzzling verbiage associated with qualified realism is as follows:

Qualified realism about F's is the thesis that F's exist, but no fundamental fact contains an F as a constituent.

When a philosopher tells you that numbers are not real in the metaphysical sense (even though they exist in the mathematical sense) one thing he or she might mean is this: that every fact in which a number figures ultimately obtains in virtue of some collection of facts in which numbers do not figure as constituents.

The idea gains support from reflection on the ontological status of 'emergent' or 'higher-level' entities in other areas. Consider *the US dollar*, or *the European Union* or *the information encoded on my hard drive*. There is a straightforward sense in which these things exist. The dollar exists as a genuine form of currency in the sense in which the Italian lira once existed but no longer does. There is a European Union in the sense in which there is not, but might one day be, an Intergalactic Federation. My hard drive really does encode a certain body of information—information that might be destroyed if I press the wrong sequence of keys. And yet there is a powerful tendency to think that while these existential claims are all perfectly correct, it would be a mistake to think of the US dollar as a Thing in the sense in which a table or God—or a number as conceived by the Platonist—would be a Thing. The proposal allows us to make sense of this tendency. It is very natural to suppose that every fact about the dollar or the EU is ultimately grounded in facts about things of a different sort—facts about the attitudes and actions of economic actors and the gold reserves in Fort Knox; facts about the legal arrangements among EU states; etc. Of course there is little hope of stating in any finite way the complete account of *that in virtue of which* the dollar is currently valued at 0.6825 euros. But the proposal does not require that. Unlike some earlier conceptions of reduction, reduction in our sense is not a thesis about the meanings of sentences in the higher-level vocabulary, or about the possibility of translating such sentences into another idiom. It is simply the claim that every fact about currencies and the like is ultimately grounded in some perhaps infinitely complex pattern of facts about other things. Now of course monetary reductionism of this sort may not be true. That is a substantive question in the philosophy of economics. The proposal is simply that our intuitive sense that the US dollar is ontologically 'derivative' or 'second-rate' derives from our strong suspicion that the dollar is unlikely to be a fundamental thing in the sense defined above.

As formulated, the proposal entails that there are bona fide Things only if there are fundamental facts. And it may be objected that we cannot assume this a priori. For all we know, there may be infinite descending chains or *trees* in which P is grounded in Q and R, which are in turn grounded in S, T, U and V, and so on ad infinitum.

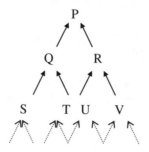

As a crude model, we might imagine a world in which facts about atoms are grounded in facts about quarks and electrons, which are in turn grounded in facts about 'hyperquarks' and 'hyperelectrons', and so on forever. Of course, the mere existence of *some* ungrounded trees would not entail that there are no fundamental things, since there might still be fundamental facts elsewhere. The worry arises only if *every* fact is grounded in further facts, in which case our definitions entail that there are no fundamental things.

Is this an objectionable consequence of our proposal? The best reason for thinking so runs as follows. Worlds in which every tree is ungrounded are of two sorts. In some of these worlds, absolutely every object that is a constituent of a fact at one level ultimately 'disappears' as we proceed down the tree, chasing each fact into the facts that ground it. In that case our proposal entails that while many things exist in the ordinary sense, nothing is ultimately real. Rather everything has the same metaphysical status as the US dollar, and that seems right. If old things dissolve and new things emerge as we boost the magnification of our metaphysical microscope, then it seems right to say that everything is on a par, metaphysically speaking, and that nothing has the ultimate status that a fundamental thing would have if there were any.

But there is also the possibility that even though every *fact* is grounded in further facts, certain things are nonetheless *persistent*, in the sense that when *x* is a constituent of some fact [...*x*...], it is also a constituent of every fact in which [...*x*...] is grounded. Intuitively (the objection runs) a persistent thing is every bit as real as a fundamental thing would be, and a thing can be persistent without being fundamental; so we should not say that a thing is real if *and only if* it is a constituent of some fundamental fact. We should say that a thing is real if and only if it is either fundamental or persistent.

In response, I suggest that the problem will not arise. Suppose that *x* is a persistent thing in the intended sense, and consider the fact that x is self-identical: [*x* = *x*]. It is totally unclear how a simple fact of this sort might be grounded in more basic facts. Since this fact is fundamental, it follows that *x* is a fundamental thing. And if this is right, we may retain our definitions, since any persistent thing will also be a fundamental thing.

Recapitulation and outstanding problems

Philosophers who contemplate mathematics often find themselves pulled in two directions. On the one hand, the spectacular success of mathematics on its own terms and in the sciences, together with the rank obviousness of some of its basic principles, inclines them to suppose that the claims of standard mathematics must be true, and hence that mathematical objects of various sorts must exist. On the other hand, the idea that mathematical objects are on a metaphysical par with paradigmatic things—we've taken concrete particulars as our examples, but we might also consider God and the angels if they exist—strikes many philosophers as preposterous. We have been struggling to formulate a position that does full justice to both of these tendencies, and we have a proposal. Qualified realism about mathematical objects is the thesis that while numbers and the like exist in the sense that sentences affirming their existence are literally true, they are not fundamental, in the sense that every fact in which the numbers figure is ultimately grounded in facts that do not involve numbers as constituents.

We may note in passing that this account explains why certain idealist and constructivist theories of mathematics have often been regarded as incompatible with full-strength realism. The mathematical idealist (to use a single label to cover a range of views) accepts that at least some of the core claims of standard mathematics are literally true, but insists that the mathematical facts are somehow grounded in our mathematical thought or activity. The crudest versions of this approach identify the objects of mathematics with ideas in the mind (but *whose* mind?). Subtler versions insist that the mathematical truths are somehow grounded in our practices, insisting that it is only because we have accepted a mathematical framework in which (e.g.) the ordinary rules of arithmetical calculation are deemed valid that it is true to say that $235 + 657 = 892$. Proponents of such views are often cagey about how it is that we, by our practices, manage to conjure the objects of mathematics into existence. It has always been obvious that we do not literally make the numbers in the sense in which a builder makes a house. But if that analogy is a dead end, what sense can we give to the idea that the numbers are somehow our creation? I will not pursue the matter except to say that our framework provides a way of stating this position. The general scheme for mathematical idealism is as follows:

For every mathematical fact [A],

[A] obtains in virtue of... (something about the thoughts, activities or practices of conscious beings).

In fact this is a good general scheme for idealism and constructivism in every area of philosophy. Berkeley's idealism about external objects might be understood as the claim that every fact involving tables and chairs is ultimately grounded in facts about the mind of God. Mill's secular idealism would substitute facts about possible human sensations. Kantian Constructivism in ethics might be the thesis that every fact about right and wrong is ultimately grounded in facts about the judgements to which any rational agent is committed when he deliberates about what to do. Social constructivism of the sort that used to be fashionable in the humanities might be the view that facts about social reality (and in the absurd limiting case, facts about reality *tout court*) are ultimately grounded in facts about our epistemic practices for accepting and rejecting claims about that reality. These views are sketchy in the extreme, but they all have the right general form: facts of one sort—facts which do not seem to concern human thought directly—are said to be grounded in facts about human thought or practice. The characteristic objects of the higher-level discourse—ordinary objects, social classes, moral obligations, etc.—all disappear when we examine the underlying facts, to be replaced by items of a different sort: representations, ideas, people and their practices, etc. These radical forms of idealism may be mistaken. But they exert a perennial (and mysterious) attraction for philosophers and philosophically minded scholars in other disciplines, so it is presumably worth seeking a clear statement of them, as well as an account of why they strike us as incompatible with an unqualified realism of the domain in question. The present framework provides the beginnings of such an account.[15]

Let us return now to the mathematical case. Our framework puts us in a position to entertain reductionist theses of the form:

For every mathematical fact [A] in some area:

[A] obtains in virtue of (...some fact that does not involve mathematical objects as constituents).

But having entertained such theses, how are we to assess them? As our examples begin to show, there are *many* ways to associate the arithmetical facts with facts in a more fundamental-seeming idiom—proof-theoretic facts, set-theoretic facts, facts of pure modal logic, etc.—in which the objects of arithmetic do not figure as constituents. These proposals are all materially adequate: they pair truths with truths; and when properly constructed, they preserve logical relations. Indeed, in many cases they respect an even more stringent (though somewhat elusive) constraint, which we might call *relevance*. As noted, any proof of an arithmetical statement A is readily converted by trivial steps into a proof of its formalist or modal structuralist counterpart, and vice versa. Speaking strictly there is a difference between the fact that there are two primes between 15 and 20 and the fact that every possible omega

---

[15] It is common to suppose that idealism and constructivism are opposed to full-strength realism because they posit the *mind-dependence* of certain apparently mind-independent facts. On the present account, this is half-right: these views are alternatives to realism because they hold that certain facts *depend* on lower-level facts involving objects of other kinds. That these lower-level facts involve *minds* is interesting, but incidental. Formalism and structuralism in the philosophy of arithmetic are also alternatives to full-strength realism, and for the same reason, and yet in these cases the underlying facts are in no way psychological.

sequence has two prime elements between its 16th and 21st elements, or the fact that the sentence 'There are two primes...' is provable in PA_ω. The facts are different because they have different constituents. And yet the claims are so close in mathematical significance that one is tempted to say that from a mathematical point of view they represent the *same* fact. I have not said this, because I do not know how to explain the notion of sameness of fact that the claim involves. (My own account is much too fine-grained to support such claims.) Still, one might say that any decent reduction of arithmetic must have this feature: the facts to which the arithmetical facts are reduced must have at least approximately the same mathematical significance as the arithmetical originals. The worry is that even if we impose this constraint, there will be *many* equally compelling reductionist proposals for arithmetic, and in general: if there is one plausible reductionist account for some area of mathematics, there will be many.[16]

This is a worry because it is natural to think these proposals cannot all be correct. If a fact about the numbers obtains in virtue of some fact about the provability of a sentence in PA_ω, it is implausible that it should *also* obtain in virtue of some quite different fact about all omega sequences. Now it should be stressed that over-determination of this sort is possible in principle. If we have a disjunctive fact [p or q], where p and q are both true, then [p or q] obtains in virtue of [p] and also in virtue of [q]. But the idea that a unified domain of fact—the arithmetical facts—might be massively and systematically overdetermined by facts from two or more rather different domains may seem implausible. And yet if we reject this possibility, the reductionist must choose. What basis could he possibly have for such a choice? It is hard to say.

One response to this predicament is a form of scepticism. If the competing reductions are incompatible, and if we have no basis for choosing among them, then surely, the only appropriate response is suspension of judgement. On this view, questions about whether and how the mathematical facts are grounded in more basic facts are fully meaningful but ultimately unanswerable. This is not an absurd idea. Why on earth should we have the resources to answer every abstruse metaphysical question that we can entertain? Still it would be good to know whether we can resist this disappointing denouement.

Here is one possible way out. So far we have considered the matter from the top down, as it were, beginning with the mathematical facts and asking for more fundamental facts in which they might be grounded. This exercise presupposes that when we consider an ordinary mathematical claim—e.g., the claim there are two primes between 15 and 20—we have a single definite fact in view, a fact whose basis we may then consider. But perhaps this is a mistake. Suppose it turns out that for each putative basis for arithmetic—the formalist basis, the modal structuralist basis, etc.—there is a distinct domain of numbers, facts about which are constituted by the basic facts in question. On this view, there is no such thing as *the* system of natural numbers. There are rather the

---

[16] For a survey of reductionist proposals for analysis, see Burgess and Rosen (1997, ch. B).

formalist numbers, facts about which are grounded in facts about provability in PA_ω, the *modal structuralist numbers,* facts about which are grounded in facts about all possible omega systems, and so on. Since the differences between these systems of numbers make no mathematical difference, the language and practice of mathematics will have had no occasion to distinguish them. If this is right, then ordinary mathematical language will be rife with *semantic indeterminacy.* When I point to a river in the distance and say, 'Let's call that river "Alph"', I introduce a meaningful word; but since I have not bothered to determine which of the many river-like objects in the vicinity I mean to pick out, my word does not denote a single thing. Rather it 'divides its reference' over a range of candidates—some a bit wider than others, some a bit longer, etc. When I use the word in a sentence, I simply have not made up my mind about which of these candidates I mean to denote. And in such cases it is misleading to speak of *the* fact that Alph is in Xanadu. Since there are many river-like objects in the vicinity, there are many such facts, the differences between which simply do not matter for my purposes. Likewise, if there are many systems of number-like objects distinguished only by the ways in which the facts about them are grounded and not by any mathematically important features, it makes no sense to speak of *the* fact that $235 + 657 = 892$. There are rather many equally qualified facts in the vicinity, each concerning numbers of some determinate kind, each grounded in some determinate way in underlying facts.[17]

If this is the metaphysico-semantic situation, then it is no wonder that we do not know how to answer questions about how the facts of arithmetic are ultimately grounded. We do not know how to answer these questions because they have a false presupposition, viz., that there is a unique system of mathematical objects and determinate range of facts about them to which the ordinary language of arithmetic manages to refer.

It is far from certain that this combination of metaphysical pluralism and semantic indeterminacy is ultimately coherent. In order to address the issue we would need a general theory of the conditions under which facts of one kind *give rise to* or *generate* facts of another kind, and the project of producing a general theory of this sort is genuinely daunting. But let us suppose for a moment that the view is not only coherent but correct. We may then note that even though questions about *the* numbers are to be rejected as badly posed, we might still endorse a form of qualified realism about arithmetic. For it might turn out that *every* candidate interpretation of the language of arithmetic takes that language to describe a class of objects that disappear upon reduction. If that is so, then it will still be true to say that arithmetical objects are not to be

---

[17] We have a precedent for this within ordinary mathematics. In certain foundational contexts it is important to distinguish the natural number 235 from the integer 235, the real number 235.0, the complex number $235 + 0i$, etc. But in many normal contexts these differences do not matter. If someone invokes *the* fact that $235 + 657 = 892$ in such a context, his speech act misfires, in the sense that he fails to pick out a single fact with his words, though the misfire may be harmless.

found among the fundamental things, and hence that the objects of arithmetic are not ultimately real.

This brings us to the hardest question that arises within this framework. Suppose that each putative reduction is associated with its own class of 'numbers', as above. This dissolves the dispute between partisans of competing reductions. But there remains an intelligible dispute between the qualified realist who says that *every* system of numbers is reducible in this way, and the full-strength realist who says that at least one system of number-like items exists at the fundamental level. In the framework we have been discussing, these are both meaningful hypotheses. And yet nothing we have said indicates a way of deciding which is right.

Some philosophers will be inclined to wield a principle of parsimony at this point. They will say that in framing an account of fundamental reality, we should aim to get by with as few things (or as few categories of things) as possible. If ground-level numbers are not needed in the philosophy of mathematics—if reducible numbers will always 'do'—then ground-level numbers are dispensable and we theorists should reject them. To proceed in this way is to assume a priori that fundamental reality is a sparsely populated realm of clear skies and desert landscapes, and speaking personally, I see no basis whatsoever for this assumption. But if parsimony cannot help us, we may find ourselves in a quandary to which the only sensible response is suspense of judgement. We may find ourselves pushed, in other words, to the view that the question of qualified realism about arithmetic is fully meaningful and yet unanswerable by any method we can imagine.

In some parts of philosophy, this sort of impasse is a sign that one's questions are badly posed. Is that so in this case? I don't think so. I believe that the grounding relation that figures in the formulation of this debate is (or can be made to be) fully intelligible. The hard question of realism about mathematics—i.e., the question whether *some* mathematical objects exist at the fundamental level—is therefore clear. I confess that I have no idea how one might go about answering it. But that is not to say that it is ultimately unanswerable. Philosophers have not always distinguished *ordinary* ontological questions—questions about what exists *simpliciter*—from *deep* ontological questions about what exists at the fundamental level (if there is one). As a result, we have no clear paradigms of inquiry into questions of the latter sort, and so no clear sense of what it takes to establish a claim about what is 'ultimately real'. Now that we have made the distinction, we can review the record, so to speak, in order to ask whether we have examples in philosophy or in physics or even in theology of arguments that bear specifically on claims about fundamental reality. If we find plausible examples, we may be able to draw explicit methodological morals from them. The way forward in the metaphysics of mathematics may then consist in bringing these methodological principles to bear on questions about the grounding of mathematical truths. Is this project likely to bear fruit? At this point, it seems to me that we have no clear basis for an answer.

Comment on Gideon Rosen's 'The reality of mathematical objects'

Timothy Gowers

At the beginning of Gideon Rosen's contribution to this volume, he describes a certain kind of philosophical position about mathematics. Ever since I myself have had considered philosophical views about mathematics, they have been of exactly that kind. Not being a professional philosopher, I have never tried to work out a fully detailed defence of my views, but I have always been confident that they are substantially correct.

If anyone is capable of shaking that kind of confidence, it is Rosen, who has a remarkable ability to articulate and then closely analyse the philosophical positions of others. He also has a way of putting forward views that I instinctively react against, such as a realism about mathematical objects, but in a way that makes them much less objectionable. For instance, if you say to Rosen that you do not believe in a metaphysical realm where numbers are floating around and enjoying various complicated relationships with each other, he will tell you that he does not either. He may then ask you why you say that there are infinitely many primes if you do not actually believe that that is the case. After a short conversation of this kind, it becomes hard (for me at least) to understand exactly what the difference is between non-realist views and sophisticated realist ones of the kind that Rosen holds.

These issues have been greatly clarified for me now that I have read Rosen's contribution to this volume. He calls the kind of position he discusses *qualified realism*: roughly speaking, qualified realism is the view that mathematical objects do indeed exist, but are, as he puts it, 'metaphysically second rate'. His aim is to say what this could possibly mean. With this aim in mind, he introduces a 'grounding relation' between facts: roughly speaking, if fact A is grounded in fact B, then fact B is more fundamental than fact A and is sufficient to explain it. (Rosen gives several different examples to illustrate and clarify this notion.) The difference between a full realist and a qualified realist, he then suggests, is that a realist believes that some mathematical facts are fundamental (that is, not grounded in any further facts), whereas a qualified realist believes that, while mathematical facts may be objectively true, they are all ultimately grounded in non-mathematical facts. As he puts it:

> When a philosopher tells you that numbers are not real in the metaphysical sense (even though they exist in the mathematical sense) one thing he or she might mean is this: that every fact in which a number figures ultimately obtains in virtue of some collection of facts in which numbers do not figure as constituents. (p. 125)

Rosen says that he does not himself endorse this view, but it resonates with me in the way that he wants it to: I think that if I were to try to put forward a full defence of my own qualified realism about numbers, I would indeed do so by trying to show that facts about numbers are grounded in other facts. There are several ways that one might go about this, just as there are several different philosophical positions that fall under the banner of qualified realism. Rosen discusses two of them in detail (neither of them the approach that I myself would follow), and also discusses the coherence of the position in general.

One of the merits of Rosen's proposal is that it replaces murky questions about the reality of numbers and mathematical statements (murky because it is often not clear what 'real' means) by the closely related but much clearer question of whether all mathematical facts are grounded in other facts. To answer this question is a large project, but it is also a clearly defined one that philosophers, and perhaps even mathematicians, could get their teeth into.

10
Getting more out of mathematics than what we put in

Mark Steiner

In *The Emperor's New Mind* (Penrose, 1989), and more recently, in *The Road to Reality* (Penrose, 2005), Professor Penrose has championed what he calls Platonism, Plato's mathematical world. Since Professor Penrose is to be present at this Symposium, I thought it would be appropriate to discuss his view, or, rather what I think his view should be, based on the thrust of his published work.

What Penrose is after, as he explains in *Road*, is the objectivity of mathematics[1]—rather than the existence of 'mathematical objects,'—whether, as another speaker at this Symposium, Professor Rosen, puts it, mathematics is a 'subject with no object'. As Burgess and Rosen point out in their excellent book (Burgess and Rosen, 1997), Platonism was hijacked by Quine, who defined it as 'quantifying over mathematical objects', and it no longer means what it used to mean. Book after book (most of them published by Oxford University Press) appear with learned discourses about, for example, whether, should 'mathematical objects' cease suddenly to exist, it would make any observable difference.[2]

I think, therefore, it would be better to eliminate reference to Plato altogether, and speak of concepts, rather than objects, and look to Descartes as the

---

[1] A modern source for the idea of emphasizing 'objectivity' over 'objects' in the philosophy of mathematics are the writings of Georg Kreisel.

[2] This question, by the way, is meaningless according to Quine's philosophy, since according to his form of Platonism we cannot describe even the observable world without 'quantifying over' mathematical objects. The indispensability of atoms is different—we can describe the appearances without appeal to atoms, but we need the atoms to explain those appearances. Quinian indispensability is, as my late and lamented teacher Sidney Morgenbesser used to say, a form of Kantianism, which Quine probably inherited from C. I. Lewis, and thus it is misleading to speak of the 'Quine–Putnam indispensability thesis'; Putnam uses indispensability in the latter sense.

real source of Penrose's feeling that such concepts as that of the Mandelbrot set are objective. After all, Penrose does not think that *every* concept is objective (as did Plato), and neither did Descartes, as he puts the matter in Meditation III:

> *...we must notice a point about ideas which do not contain true and immutable natures but are merely ones which are invented and put together by the intellect. Such ideas can always be split up by the same intellect, not simply by an abstraction but by a clear and distinct intellectual operation, so that any ideas which the intellect cannot split up in this way were clearly not put together by the intellect. When, for example, I think of a winged horse or an actually existing lion...I readily understand that I am also able to think of a horse without wings, or a lion which does not exist...and so on; hence these things do not have true and immutable natures. But if I think of a triangle or a square...then whatever I apprehend as being contained in the idea of a triangle—for example that its three angles are equal to two right angles—I can with truth assert of the triangle...*

What Descartes is saying here is that you get more out of mathematics than what you put in; there is 'latent information' inherent in mathematical ideas that is not contained in their verbal definition.[3] This 'latent information' is part of the essence of these ideas, not put there by any mathematician. This is, in my opinion, the real difference between mathematics and games. Take the following position in a chess endgame (see Fig. 10.1).[4]

White needs to play 262 accurate moves to mate, assuming perfect play on both sides. In many cases the moves 'make no sense', in the sense that one cannot explain them without actually giving the entire tree of moves. Though obviously this kind of thing is not 'expected' (the 50 rule move, which is violated five times in the winning 'line', proves this), the surprise here is how *little* is contained in the rules of chess, how little you reap for the investment. Thus, the essence of chess is arbitrary, certainly not one of Descartes' 'true and immutable essences'.

Compare this to one of Professor Penrose's favorite examples, the 'magical' complex numbers. When they were introduced by the Italians as imaginary solutions for equations over the reals, nobody could have predicted that they would play the role of tying together the real numbers, as in the beautiful equation

$$e^{\pi i} + 1 = 0$$

---

[3] I think Descartes is saying more than just that some mathematical truths are *synthetic* in the sense of Kant. For Kant, both $7 + 5 = 12$ and the theorem about the sum of the angles of a triangle are synthetic truths, but in the first case we don't learn anything of mathematical interest, since the sum had to be either one number or another. The sum of the angles of a triangle is real information.

[4] See http://www.chessbase.com/newsroom2.asp?id=239; thanks to Sylvain Cappell for supplying this URL.

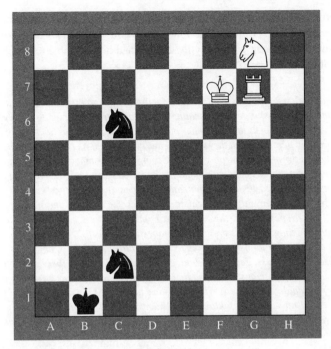

Fig. 10.1 White to play and mate in 262 moves.

which is a special case of Euler's discovery

$$e^{i\theta} = \cos(\theta) + i\sin(\theta)$$

Note that when imaginaries were introduced, the idea of raising a real number to an imaginary power was unthinkable even to Cardano and Bombelli. Yet once one thinks of the idea, it turns out that there is little or no choice in how to proceed.

Another property not recognized by the originators of imaginary numbers is the *absolute value* of a complex number, which emerges when we embed the complex numbers in the Euclidean plane. Using this property, mathematicians can explain facts about the real numbers; for example, why the real function $\frac{1}{1+x^2}$, defined everywhere on the reals, is not equal to its expansion as a power series $1 - x^2 + x^4 - \cdots$ wherever $|x| \geq 1$ (the complex numbers with absolute value 1 form a circle around the origin, and the real number 1 and the imaginary i lie on this circle. For i, the function, even when continued into the complex plane, is not defined, since the denominator is zero. Standard theorems in complex analysis do the rest).

One might say that the introduction of imaginaries is motivated in part by computational convenience: Cardano uses imaginary numbers even in computing real roots of a cubic equation (which can appear in his famous formula).

Even so, the 'latent information' inherent in his idea goes far beyond calculational convenience. All in all, the following passage from *The Emperor's New Mind* could have been written by Descartes (had he been able to prophesy the future of mathematics):

> *While at first it may seem that the introduction of such square roots of negative numbers was just a device—a mathematical invention designed to achieve a specific purpose—it later becomes clear that these objects are achieving far more than that for which they were originally designed. As I mentioned above, although the original purpose of introducing complex numbers was to enable square roots to be taken with impunity, by introducing such numbers we find that we get, as a bonus, the potentiality for taking any other kind of root or for solving any algebraic equation whatever. Later we find many other magical properties that these complex numbers possess, properties that we had no inkling about at first. These properties are just there. They were not put there by Cardano, nor by Bombelli, nor Wallis, nor Coates, nor Euler, nor Wessel, nor Gauss, despite the undoubted farsightedness of these, and other, great mathematicians; such magic was inherent in the very structure that they gradually uncovered. When Cardano introduced his complex numbers, he could have had no inkling of the many magical properties which were to follow—properties which go under various names, such as the Cauchy integral formula, the Riemann mapping theorem, the Lewy extension property. These, and many other remarkable facts, are properties of the very numbers, with no additional modifications whatever, that Cardano had first encountered in about 1539.*
>
> Penrose (1989, pp. 96–97)

The Descartes/Penrose idea of 'true and immutable essences' in mathematical concepts, which I discussed in a different essay some years back (Steiner, 2000), has nothing particularly to do with applications of mathematics to nature. Nevertheless, and this is the thesis of the present essay, it quite often turns out that the very latent mathematical information found in mathematical concepts—even those introduced for 'computational convenience'—provide the most spectacular applications of mathematics in natural science. This 'surplus value' is particularly glaring in the application of 'imaginary' numbers to 'real' nature. As in pure mathematics, the origin of these applications is in calculational convenience, yet they end up in descriptive necessity. In what follows, I will discuss some very well known facts (to those who have studied Penrose's works), so well known that we may forget just how remarkable they are. From now on, we will be discussing what is called the 'unreasonable effectiveness' of mathematics in natural science.

Euler's discovery makes for a convenient way to represent rotations in the plane without messy trigonometric formulas—namely, by a unit vector in the complex plane. Composition of two rotations is given by multiplying two unit vectors, giving a unit vector whose argument (angle) is the sum of the two arguments.

The 19th century saw a number of attempts to generalize this convenience to rotations in space. Euler had shown how to represent spatial rotations by three angles (called today the 'Euler angles'); it seemed reasonable to generalize the notion of complex numbers to three-dimensional vectors in a 'complex space'. These attempts failed, and Hamilton was forced to ascend to the fourth dimension to get a vector space endowed with a multiplication, in analogy to the complex numbers.[5] The elements of this algebra he called 'quaternions' of the form $a + bi + cj + dk$ where the multiplication of the alignment elements were governed by $i^2 = j^2 = k^2 = ijk = -1$, an equation he carved into a bridge in his excitement on finding these relations. In analogy to complex numbers, the *unit* quaternions represent spatial rotations. Multiplying two unit quaternions represents the composition of the two represented rotations, so we have (using a later terminology) a homomorphism. The fact that quaternion multiplication is not commutative is fine for this purpose, since the rotations themselves are not commutative. There are, however, some puzzles: the special cases of i, j, k represent rotations about the x, y, z axes, respectively—only rotations of 180°, not 90°, otherwise the successive rotations represented by i, j, k will not bring the axes back to their original alignment. (See Penrose, 2005, ch. 11, for elucidations.) This means that the three unit quaternions $-i, -j, -k$ all represent rotations of 540°(= 180°) as well. This is true in general—the negative of a unit quaternion represents the same rotation as the quarternion itself. The homomorphism is not an isomorphism, but two to one. Any continuous (sub)group $r(\theta)$ of spatial rotations about a fixed axis can be represented homomorphically by a continuous path $q(\theta)$ of unit quaternions such that $q(\theta + 2\pi) = -q(\theta)$, while $r(\theta + 2\pi) = r(\theta)$. Each rotation gets two labels, or 'parities'. The parity of a rotation is useless information, or so it seemed.

Another method of extending the representation of plane rotations by complex numbers of magnitude one to spatial rotations evolved at roughly the same time. This was representation of rotations by 2×2 unitary matrices of (ordinary) complex numbers. A unitary matrix $\mathbf{M}$ satisfies $\mathbf{MM^*} = \mathbf{I}$ (the identity matrix), where $\mathbf{M^*}$ is the matrix we get by transposing rows and columns in $\mathbf{M}$, and also replacing all four entries with their complex conjugates (i.e., $x + iy \rightarrow x - iy$). The determinant of a unitary matrix must be a complex number with absolute value equal to unity, and restricting the matrices to those with determinant $+1$, we arrive at what is today called SU(2), the 'special unitary' group of 2×2 matrices. This method is the fruit of the efforts of mathematicians like Cayley, Laguerre and others. But the earliest I could find

---

[5] I am grateful to Sylvain Cappell and Stanley Ocken for improving the formulation here. Professor Cappell brought to my attention a deep result of Frank Adams, the English topologist, that only in dimension 1, 2, 4 and 8 does Euclidean space have a vector space with a multiplication that satisfies: (a) there is a multiplicative identity (left and right); (b) the product of two non-zero vectors is non-zero. For n = 8, the multiplication is not associative.

explicitly the homomorphism between SU(2), and the rotation group is 1884, when Felix Klein (1888, p. 34ff)[6] lectured on the icosahedron group and two years later used the homomorphism to calculate with rotations in his study of the gyroscope (see Klein, 1922).

SU(2) turns out to be isomorphic to the unit quaternions, so there is also a two-to-one homomorphism from SU(2) onto all the rotations. For example, the SU(2) matrices

$$\begin{pmatrix} e^{i\theta} & 0 \\ 0 & e^{-i\theta} \end{pmatrix}$$

correspond to a rotation of 2θ around the z-axis[7] When $\theta = 0$, the matrix is the identity matrix $I = \begin{pmatrix} 1 & 0 \\ 0 & 1 \end{pmatrix}$ corresponding to the null rotation; when θ reaches π, the matrix is $-I$, and the rotation is a full rotation. It takes two full rotations to bring the matrix back to I. Again, we have a kind of gratuitous labelling of the rotations as positive or negative, rotations in the interval $(360°, 720°)$ being with SU(2), and calculational convenience was the rationale of introducing SU(2) in the first place.

What I find remarkable is that this very superfluous information turns out to be the key to some of the fundamental features of our universe. For the symmetry of the electron turned out precisely to be that of SU(2). When we rotate the electron 360° its quantum mechanical description (wave function) is multiplied by -1! To get the electron back to its initial state, one must give it two full turns. Had mathematical physicists been much better calculators, they might have missed one of the spectacular discoveries in the history of science.[8] This is one of many examples in which human limitations, far from impeding scientific progress, were responsible for it.[9]

---

[6] Klein does *not* explictly state that spatial rotations can be represented by 3×3 real orthogonal matrices with determinant $+1$, so we cannot yet credit him with the discovery that SU(2) is two-to-one homomorphic onto SO(3).

[7] We set up another correspondence between self-adjoint matrices and 3 vectors in space (see Sternberg (1994, §1.2) or Goldstein (1980, ch. 4) for details). Let **A** be a self-adjoint matrix, and **M** a member of SU(2). Then conjugation by **M**, i.e., **MAM***, yields a self-adjoint matrix corresponding to a rotated vector.

[8] See also Hadamard (1954, pp. 128–189): 'Most surprising—I should say bewildering—facts of that kind are connected with the extraordinary marks of contemporary physics. In 1913, Elie Cartan, one of the first among French mathematicians, thought of a remarkable class of analytic and geometric transformations in relation to the theory of groups. No reason was seen, at that time, for special consideration of those transformations except just their esthetic character. Then, some fifteen years later, experiments revealed to physicists some extraordinary phenomena concerning electrons, which they could only understand by the help of Cartan's ideas of 1913.'

[9] The limitations on observation made it impossible for Kepler to observe the perturbations of one planet by another, and he published his 'laws' of planetary motion as though they were not so perturbed. This enabled Newton to derive from these laws

(a) (b)

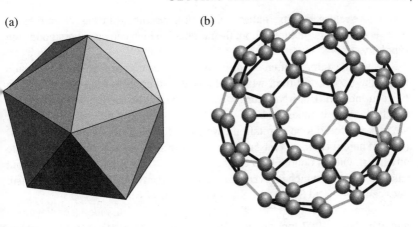

Fig. 10.2 (a) Icosahedron. (b) Carbon-60 molecule (buckyball).

The fact that the electron 'lives' in a two-dimensional complex vector space and has SU(2) symmetry is detectable even on the macroscopic level. Consider the carbon-60 molecule, which has the shape of a buckyball, or the truncated icosahedron, which is made out of hexagons and pentagons like a soccer ball. The buckyball would appear to have the symmetry I of an icosahedron: sixty different rotations around its center leave the buckyball invariant.[10] Yet if we want to study the paramagnetic behaviour of the C-60 molecule in a magnetic field that also has icosahedral symmetry we must 'pull back' to the members of SU(2) that correspond to the members of I: the true symmetry group of the buckyball, then, is the 120 member subgroup, G, of SU(2) that is two-to-one homomorphic onto I (Chung *et al.*, 1994, §9).

There is a story within a story here. Laguerre published in 1867 a 'letter' to Hermite in which he defines matrix multiplication and discusses its properties (Laguerre, 1867: 1898). In the course of this discussion, he introduces matrices over the integers modulo p, presumably for studies in number theory, but in any case, not for any physical application. If we consider 2×2 matrices, let $p = 5$, and restrict ourselves to invertible matrices whose determinant is $+1$, we get exactly 120 matrices forming a group, and this group turns out to be isomorphic to G, a rather startling fact, which means that Laguerre's number-theoretic idea turns out to describe the electronic and magnetic properties of the buckyball—surely we got more out than we put in (Chung *et al.*, 1994, §2)!

Let us now return to the electron itself, which returns to itself only after two turns. It might be thought that, since the electron ignores standard geometry, it

---

the inverse square law of gravitational force, and *then* study the resulting perturbations. The 'missing' lines in the spectrum of hydrogen were actually there (though weak); had they been detectable (then), a major discovery would have been missed.

[10] See Chung and Sternberg (1993) for a nice exposition.

could have returned to itself after 3 or any other number of turns. Nevertheless, there is something 'natural' about the number 2 which can be understood using topology.

Using an example often cited, if you place the end of a belt in a book and rotate the book a full turn, while holding the other end of the belt, you get a twist in the belt. But it you turn the book two full turns, you can untwist the belt by looping it under the book. Topologically this is expressed as follows: if we make a closed loop of rotations, we cannot 'shrink' this loop into a 'point' (a point here meaning a constant curve of rotations, a curve that doesn't rotate anything). Only a double loop can be so shrunk. This is an indirect fact about Euclidean space, which is revealed by studying the two-to-one homomorphism between SU(2) and SO(3) (see Sternberg, 1994, §1.6).

Feynman has another "twist" (excuse the pun) on this idea (see Feynman and Weinberg, 1987, pp. 56–59). He suggests tying a belt to two electrons, and get them to switch places. The belt will be found to be twisted, showing that topologically we have a single rotation, each electron having made what is equivalent to half a rotation. In that case we should expect the wave function describing the two electrons to change signs, which is the fundamental property of fermions. If this argument is true, and it seems too good to be true, we get the famous connection between 'spin' and 'statistics' using SU(2) and a little topology—no relativity, no field theory. Even if this argument is inconclusive, there *is* a proof connecting SU(2) symmetry with fermionic statistics using considerations that the inventors of SU(2) could not have dreamed of.

What is most remarkable, perhaps, is the further application of SU(2) and unitary matrices in general. These further applications disconnect the concept of unitary matrix more and more from their original application—the representation of rotations.

I will cite two examples: the first is the application by Heisenberg to nuclear physics in 1932 of SU(2) symmetry. It turned out that the neutron and proton are two states of the same particle (called today the 'nucleon'), and that, mathematically, the two particles are analogous to the 'spin up' and 'spin down' states of the electron. The 'rotation' that would take the neutron into the proton and then back to the neutron (a 'rotation' that would cause the wave function of the particle to change sign) cannot be thought of as a rotation in physical space; nevertheless, the SU(2) symmetry of the nucleon, called 'isotopic spin', has empirical consequences. The physical basis for this analogy, if there is one, is unknown even today.

When we move to higher-dimensional unitary matrices, the simple connection with rotations breaks down. For example, SU(3) (the group of 3×3 unitary matrices with determinant $+1$), is not homomorphic onto any group SO(n). Its connection to rotation is only via the analogy to SU(2). Despite this, or perhaps because of this, SU(3) turns out to describe the three states of a (type of) quark, which the nucleons are made of (since the nucleons have integral charge, the quarks have fractional charge, a fact which impeded their discovery). Rotations turned out to be the tip of a much bigger iceberg.

Another way to generalize the SU(2) two-to-one homomorphism onto the rotation group was noted by Klein in his lectures on the top (1922, Lecture I). If we drop the requirement of unitariness, and consider only 2×2 complex matrices with determinant 1, we get the group of matrices today known as SL(2,C). Klein shows (see p. 626 for a statement of this theorem) that there is a two-to-one homomorphism between SL(2,C) and a group of transformations on a two-dimensional manifold, and this group turns out to be isomorphic to the proper Lorentz group L° (the group of Lorentz transformations that have positive determinants and preserve the forward 'light cone'), a homomorphism which has itself great physical significance when we are studying relativistic quantum theory.

I think that the applicability of 'magical' (Penrose's phrase) complex numbers to rotations and its generalization to SU(2) and beyond is not unusual in the history of mathematics, but it well illustrates the phenomenon that mathematical concepts tend to have latent information, which can be used in the development of mathematics. What is more, nature seems to make use of these mathematical possibilities.

There are three elements coming together here: mathematics, nature, and the human mind. Which of these three is responsible for the remarkable richness of 'mathematical essences'?

Acknowledgements

I would like to thank Shlomo Sternberg for crucial aid and advice. Thanks also to Shmuel Elitzur, Sylvain Cappell, and Carl Posy for valuable discussions over the course of many years. The research that went into this article was supported by the Israel Science Foundation, Grant no. 251/06, and I am very grateful for this support.

Comment on Mark Steiner's 'Getting more out of mathematics than what we put in'

Marcus du Sautoy

The idea of getting more out than you put in is one of the most attractive features of my subject. I think this quotient is probably higher for mathematics than for any of the other sciences, which is one reason I chose mathematics over the mess of biology. Maybe it's also because I have such a bad memory. You only need a few axioms and everything else starts spilling out once you start. Biology and chemistry always seemed to require you to memorize the things Nature selected, which at times appear quite arbitrary.

The art of the mathematician is often to pick the rules of the game to maximize this quotient. Although Steiner makes a distinction between games and maths, I think there is the same satisfaction in seeing how the simple rules of 'Go' lead to a game with such a rich array of plays, and the way three simple axioms for the definition of a group can lead to the Monster simple group and all the Lie groups that underpin physics.

Which brings me to the challenging point raised by Steiner's article: the extraordinary synergy between the abstract, beautiful world of mathematics and the physical, messy world of physics, chemistry and biology. It is perhaps bizarre that a perfect circle or right-angled triangle, the most fundamental objects of mathematics, may not have any physical reality if quantum physics is right about the world being quantized into discrete bits. And the infinities that I bandy around with ease might have no concrete realization in a universe that is potentially finite in nature. Nevertheless it is astounding how these objects of the mind help us to predict the future behaviour of our messy universe. How come the imaginary numbers we create to solve a cubic equation are the same numbers that are crucial to describing the quantum world?

Perhaps it is an anthropomorphic answer as Steiner suggests in response to my article. We make choices about the maths we like to celebrate. Where did that maths and the excitement come from originally? From describing the physical world. The Egyptians wanted to know the volume of a pyramid. They needed to know after all how many bricks to use. But to calculate the volume they are led to the discovery of the power of cutting a shape into infinitely many, infinitely thin pieces, which they can rearrange to make the problem easier. An early form of integral calculus at work.

The process of cutting a real pyramid like this is clearly absurd on a practical level, yet a projection has been established from the world of mathematics down onto our messy world. But because the world of mathematics began it's journey trying to describe and predict physical reality, perhaps it isn't so unexpected that the maths we generate in a purely abstract form, and for its intrinsically internal fascination, nevertheless can often find itself being projected back down to our messy universe generations after the journey was kicked off.

A last point. Sometimes maths is very good at showing why you can't get any more out from what you put in. Real numbers led to complex numbers led to quaternions and gave birth to octonions, but then mathematicians can prove that you're not going to get any more out of this. Similarly, the Lie groups E_6, E_7 and E_8 are such beautifully powerful structures, but the maths shows why it stops there. There can't be an E_9. Sometimes you get less out than you might expect. But knowing that is sometimes as exciting as getting lots out of a small investment. The exceptional Lie groups are special because of their unique character. Still, it is amazing that E_8 could be the model for the fundamental particles that make up the fabric of reality. Nature certainly has good taste.

References

Archimedes. *The Method*. In *Greek Mathematical Works. II: From Aristarchus to Pappus*, ed. J. Heiberg and trans. I. Thomas, Loeb Classical Library, 362. Cambridge, MA: Harvard University Press (1993), pp. 221–223.

Arnauld, A. (1964). *The Art of Thinking: Port Royal Logic*, trans. J. Dickoff and P. James. Indianapolis, IL: Bobbs-Merrill.

Benacerraf, P. (1965). What numbers could not be. *Philosophical Review*, **74**.

Benacerraf, P. (1973). Mathematical truth. *Journal of Philosophy*, **70**.

Bohm, D. and Hiley, B. J. (1993). *The Undivided Universe*. London: Routledge.

Bolzano, B. (1810). *Contributions to a Better-Grounded Presentation of Mathematics*. In *The Mathematical Works of Bernard Bolzano*, ed. and trans. S. Russ. Oxford: Oxford University Press (2004).

Borges, J. L. (1941). The Library of Babel (La Biblioteca de Babel). In *Labyrinths*. Harmondsworth: Penguin (1970).

Burgess, J. P. (1983). Why I am not a nominalist. *Notre Dame Journal of Formal Logic*, **24**: 1.

Burgess, J. and Rosen, G. (1997). *A Subject with no Object: Strategies for Nominalistic Interpretation of Mathematics*. Oxford: Oxford University Press.

Changeux, J.-P. and Connes, A. (1995). *Conversations on Mind, Matter and Mathematics*, ed. and trans. M. B. DeBevoise. Princeton, NJ: Princeton University Press.

Chung, F. and Sternberg, S. (1993). Mathematics and the buckyball. *American Scientist*, **81**: 56–71.

Chung, F., Kostant, B., and Sternberg, S. (1994). Groups and the buckyball. In *Lie theory and Geometry: In Honor of Bertram Kostant*, ed. J.-L. Brylinski, R. Brylinski, V. Guillemin and V. Kac. Boston: Birkhäuser.

Cicero, M. T. *The Orations of Marcus Tullius Cicero*. 4 vols. London: G. Bell & Sons (1894–1903).

Coffa, A. (1986). From geometry to tolerance: sources of conventionalism in nineteenth-century geometry. In *From Quarks to Quasars: Philosophical Problems of Modern Physics*. University of Pittsburgh Series, 7. Pittsburgh, PA: Pittsburgh University Press, 3–70.

Cohen, P. J. (1966). *Set Theory and the Continuum Hypothesis.* New York: W. A. Benjamin, Inc.

Courant, R. and Robbins, H. (1947). *What is Mathematics? An Elementary Approach to Ideas and Methods.* Oxford: Oxford University Press (1981).

Curry, H. (1951). *Outlines of a Formalist Philosophy of Mathematics.* Amsterdam: North-Holland.

Dedekind, R. (1888). *Was sind und was sollen die Zahlen.* In *Gesammelte Mathematische Werke, III.* Braunschweig: Friedrich Vieweg und Sohn.

Detlefsen, M. (2005) Formalism. In *The Oxford Handbook of Philosophy of Mathematics and Logic,* ed. S. Shapiro. Oxford: Oxford University Press, pp. 236–317.

Dorr, C. (2008). There are no abstract objects. In *Contemporary Debates in Metaphysics,* ed. T. Sider, J. Hawthorne and D. W. Zimmerman. Oxford: Wiley-Blackwell.

du Sautoy, M. (2008). *Finding Moonshine: A Mathematician's Journey Through Symmetry.* London: Harper Perennial (2009).

Dummett, M. (1978). *Truth and Other Enigmas.* Cambridge, MA: Harvard University Press.

Feynman, R. P. and Weinberg, S. (1987). *Elementary Particles and the Laws of Physics.* Cambridge: Cambridge University Press.

Field, H. (1980). *Science Without Numbers.* Princeton, NJ: Princeton University Press.

Field, H. (1984). Is mathematical knowledge just logical knowledge? Reprinted with a postscript in *Realism, Mathematics, and Modality.* Oxford: Blackwell (1989), pp. 79–124.

Field, H. (1991). Metalogic and modality. *Philosophical Studies,* **62**(1): 1–22.

Frege, G. (1884). *The Foundations of Arithmetic: A Logico-Mathematical Enquiry into the Concept of Number,* trans. J. L. Austin, 2nd edn., New York: Harper (1960); 2nd rev. edn., Evanston, IL: Northwestern University Press (1968).

Frege, G. (1893). *Grundgesetze der Arithmetik: begriffsschriftlich abgeleitet,* Vol. I. Hildesheim: G. Olms Verlag (1962).

Frege, G. (1903). *Grundgesetze der Arithmetik: begriffsschriftlich abgeleitet,* Vol. II. Hildesheim: G. Olms Verlag (1962).

Gabriel, G. *et al.,* eds. (1980). *Gottlob Frege: Philosophical and Mathematical Correspondence.* Chicago: University of Chicago Press.

Gödel, K. (1947). What is Cantor's continuum problem? Revised and expanded version in *Kurt Gödel: Collected Works,* Vol. II. Oxford: Oxford University Press (1990).

Gödel, K. (1951). Some basic theorems on the foundations of mathematics and their implications. In *Kurt Gödel: Collected Works,* Vol. III. Oxford: Oxford University Press (1995).

Goldstein, H. (1980). *Classical Mechanics,* 2nd edn. Reading, UK: Addison-Wesley.

Gregory, R. (1969). *Eye and Brain: The Psychology of Seeing.* New York: McGraw Hill.

Hadamard, J. (1954). *The Psychology of Invention in the Mathematical Field.* New York: Dover.

Hardy, G. H. (1940). *A Mathematician's Apology.* Cambridge: Cambridge University Press (1967).

Hartle, J. B. (2003). *Gravity: An Introduction to Einstein's General Relativity.* San Francisco: Addison-Wesley.

Hellman, G. (1989). *Mathematics Without Numbers.* Oxford: Oxford University Press.

Herschel, J. (1841). Review of Whewell's *History of the Inductive Sciences and Philosophy of the Inductive Sciences. Quarterly Review,* **68**: 177–238.

Heyting, A. (1931). The intuitionistic foundations of mathematics. In *Philosophy of Mathematics,* 2nd edn., ed. P. Benacerraf and H. Putnam. Cambridge: Cambridge University Press (1983), pp. 52–61.

Hilbert, D. (1899). Die grundlagen der geometrie. In *Festschrift zur Feier der Enthullung des Gauss-Weber Denkmals in Göttingen.* Leipzig: Teubner.

Horgan, T. (1994). Transvaluationism: a Dionysian approach to vagueness. *The Southern Journal of Philosophy,* Supplement, **33**: 97–126.

Hutton, C. (1795–1796). *A Mathematical and Philosophical Dictionary.* 2 vols. London: J. Johnson, and G. G. and J. Robinson. Reprinted, Hildesheim and New York: G. Olms Verlag (1973), and in 4 vols., Bristol: Thoemmes Press (2000).

Kant, I. (1781). *Kritik der Reinen Vernunft,* ed. R. Schmidt. Hamburg: Felix Meiner Verlag (1990).

Kitcher, P. (1989). Explanatory unification and the causal structure of the world. In *Scientific Explanation,* ed. P. Kitcher and W. Salmon. Minneapolis, MI: University of Minnesota Press, pp. 410–505.

Klein, F. (1888). *Lectures on the Ikosahedron and the Solution of Equations of the Fifth Degree.* London: Trübner & Co.

Klein, F. (1922). The mathematical theory of the top (1896/97). In *Felix Klein Gesmmelte Mathematische Abhandlungen,* ed. R. Fricke and H. Vermeil. Berlin: Springer.

Kreisel, G. (1967). Informal rigour and completeness proofs. In *Problems in the Philosophy of Mathematics,* ed. I. Lakatos. Amsterdam: North-Holland, pp. 138–186.

Laguerre, E. N. (1898). Sur le calcul des systemes linéaires. In *Oeuvres de Laguerre,* ed. C. Hermite, H. Poincaré and R. Eugene. Paris: Gauthier-Villars.

Lakatos, I. (1976). *Proofs and Refutations,* ed. J. Worrall and E. Zahar. Cambridge: Cambridge University Press.

Leibniz, G. W. F. *Die philosophischen Schriften von Gottfried Wilhelm Leibniz,* ed. C. J. Gerhardt and C. I. Gerhardt. Hildesheim: G. Olms Verlag (1978).

Leibniz, G. W. F. *Discourse on Metaphysics and Other Essays,* ed. D. Garber and R. Ariew. Indianapolis, IL: Hackett (1989).

Leibniz, G. W. F. (1764). *New Essays Concerning Human Understanding*, trans. A. G. Langley. Chicago: Open Court (1916).

Leibniz, G. W. F. (1683). Of Universal Analysis and Synthesis; or, of the Art of Discovery and of Judgement. In *Philosophical Writings [of] Leibniz*, ed. and trans. M. Morris and G. H. R. Parkinson. London: J. M. Dent and Sons (1973), pp. 10–17.

Leibniz, G. W. F. *Opera philosophica quae extant latina, gallica, germanica omnia*, ed. J. E. Erdmann. Aalen: Scientia (1959).

Leibniz, G. W. F. *Opuscules et fragments inédits de Leibniz: extraits des manuscrits de la bibliothèque royale de Hanover*. Paris (1903).

Leslie, J. (1809). *Elements of Geometry, Geometrical Analysis, and Plane Trigonometry: with an appendix, notes and illustrations*. Edinburgh: Brown and Crombie.

Leslie, J. (1821). *Geometrical Analysis, and Geometry of Curved Lines: being volume second of a course of mathematics, and designed as an introduction to the study of natural philosophy*. Edinburgh: W. & C. Tait, and London: Longman, Hurst, Rees, Orme, & Brown.

Lipton, P. (1991). *Inference to the Best Explanation*, 2nd edn. New York: Routledge Publishing Company (2004).

Maddy, P. (2007). *Second Philosophy: A Naturalistic Method*. Oxford: Oxford University Press.

Mates, B. (1986). *The Philosophy of Leibniz: Metaphysics and Language*. Oxford: Oxford University Press.

Nelson, E. (1986). *Predicative Arithmetic*. Princeton, NJ: Princeton University Press.

Penrose, R. (1989). *The Emperor's New Mind*. Oxford: Oxford University Press.

Penrose, R. (1994). *Shadows of the Mind: An Approach to the Missing Science of Consciousness*. Oxford: Oxford University Press.

Penrose, R. (1997). On understanding understanding. *International Studies in the Philososophy of Science*, **11:** 7–20.

Penrose, R. (2004). *The Road to Reality: A Complete Guide to the Laws of the Universe*. London: Jonathan Cape and New York: Alfred Knopf (2005).

Penrose, R. (2005). *The Road to Reality: A Complete Guide to the Laws of the Universe*. New York: Alfred Knopf.

Penrose, R. (2011). Gödel, the mind, and the laws of physics. In *Kurt Gödel and the Foundations of Mathematics: Horizons of Truth*, ed. M. Baaz, C. H. Papadimitriou, D. S. Scott, H. Putnam and C. L. Harper, Jr. Cambridge: Cambridge University Press, forthcoming.

Playfair, J. (1778). On the arithmetic of impossible quantities. *Philosophical Transactions of the Royal Society of London*, **68**: 318–343.

Polkinghorne, J. C. (1996). *Beyond Science*. Cambridge: Cambridge University Press.

Polkinghorne, J. C. (1998). *Belief in God in an Age of Science*. New Haven, CT: Yale University Press.

Polkinghorne, J. C. (2005). *Exploring Reality*. London: SPCK and New Haven, CT: Yale University Press.

Proclus. *A Commentary on the First Book of Euclid's Elements*, trans. G. R. Morrow. Princeton, NJ: Princeton University Press (1970).

Putnam, H. (1967). Mathematics without foundations. *Journal of Philosophy*, **64**.

Putnam, H. (1975). What is mathematical truth? In *Mathematics, Matter and Method*, 2nd edn. Vol. 1 of *Philosophical Papers*. Cambridge: Cambridge University Press (1979), pp. 60–78.

Quine, W. V. (1960). *Word and Object*. Cambridge, MA: The MIT Press.

Resnik, M. (1980). *Frege and the Philosophy of Mathematics*. Ithaca, NY: Cornell University Press.

Resnik, M. (1997). *Mathematics as a Science of Patterns*. Oxford: Oxford University Press.

Rosen, G. (1994). Objectivity and modern idealism. In *Philosophy in Mind*, ed. J. O'Leary-Hawthorne and M. Michael. Dordrecht: Kluwer.

Rosen, G. (2006). Review of Jody Azzouni, deflating existential consequence. *Journal of Philosophy*, **103**: 6.

Rosen, G. (2010). Metaphysical dependence: reduction and grounding. In *Modality: Metaphysics, Logic and Epistemology*, ed. B. Hale and A. Hoffmann. Oxford: Oxford University Press.

Rosen, G. and Burgess, J. P. (2005). Nominalism reconsidered. In *Oxford Handbook of Philosophy of Mathematics and Logic*, ed. S. Shapiro. Oxford: Oxford University Press.

Russell, B. (1905). On denoting. *Mind*, **14**: 56.

Salmon, W. (1990). *Four Decades of Scientific Explanation*. Minneapolis, MI: University of Minnesota Press, and Pittsburgh, PA: Pittsburgh University Press (2006).

Schopenhauer, A. *Arthur Schopenhauer's Sämtliche Werke*, Vol. 2. Munich: R. Piper & Co. Verlag (1911).

Schopenhauer, A. (1859). *The World as Will and Representation* (*Die Welt als Wille und Vorstellung*). New York: Dover Publications (1966).

Shapiro, S. (1997). *Philosophy of Mathematics: Structure and Ontology*. Oxford: Oxford University Press.

Shapiro, S. (2000). The status of logic. In *New Essays on the A Priori*, ed. P. Boghossian and C. Peacocke, Oxford: Oxford University Press, pp. 333–366; reprinted (in part) as 'Quine on Logic', in *Logica Yearbook 1999*, ed. T. Childers, Prague: Czech Academy Publishing House, pp. 11–21.

Shapiro, S. (2007). The objectivity of mathematics. *Synthese*, **156**: 337–381.

Shapiro, S. (2007a). *Vagueness in Context*. Oxford: Oxford University Press.

Shapiro, S. (2009). We hold these truths to be self-evident: but what do we mean by that? *Review of Symbolic Logic*, **2**: 175–207.

Shapiro, S., ed. (2005). *The Oxford Handbook of Philosophy of Mathematics and Logic*. Oxford: Oxford University Press.

Steiner, M. (1978). Mathematical explanation and scientific knowledge. *Nous*, **12**: 17–28.

Steiner, M. (1980). Mathematical explanation. *Philosophical Studies*, **34**: 135–152.

Steiner, M. (1998). *The Applicability of Mathematics as a Philosophical Problem*. Cambridge, MA: Harvard Uiniversity Press.

Steiner, M. (2000). Penrose and Platonism. In *The Growth of Mathematical Knowledge*, ed. E. Grosholz and H. Breger. Dordrecht and Boston: Kluwer.

Sternberg, S. (1994). *Group Theory and Physics*. Cambridge: Cambridge University Press.

Tait, P. G. (1866). Sir William Rowan Hamilton. *North British Review*, **14**: 37–74.

Waismann, F. (1979). *Ludwig Wittgenstein and the Vienna Circle*. London: Blackwell.

Waismann, F. (1982). *Lectures on the Philosophy of Mathematics*. Amsterdam: Rodopi.

Weyl, H. (1921). Über die neue grundlagenkrise der mathematik. *Mathematische Zeitschrift*, **10**: 39–79.

Wittgenstein, L. (1953). *Philosophical Investigations*, trans. G. E. M. Anscombe, 3rd edn. Oxford: Blackwell (2001).

Wittgenstein, L. (1956). *Remarks on the Foundations of Mathematics*, 3rd edn. Oxford: Blackwell (1978); rev. edn., ed. G. H. von Wright, R. Rhees and G. E. M. Anscombe, Cambridge, MA: The MIT Press (1967).

Woodward, J. (2009). Scientific explanation. *Stanford Internet Encyclopedia of Philosophy*, http://plato.stanford.edu/entries/scientific-explanation.

Wright, C. (1992). *Truth and Objectivity*. Cambridge, MA: Harvard University Press.

Index

Note: page numbers in *italics* refer to Figures.

632 symmetry group 18, *19*

A₅ symmetry group 18
abductive reasoning 65
abstract artefacts 15
abstract concepts, independent existence 5
acceptance 80–1
Adams, Frank 139
aesthetic discourse, objectivity 109
aesthetics 26
  Islamic art 34
  music 33
  role in mathematical creation 20, 22
aisthēsis 73
Alhambra Palace, symmetrical designs 18, *19*,
  34
Ames, Adelbert, 'Distorted Room' 79
anagrams 4, 6
analogy
  argument by 29–30, 36
  between forcedness and sensory
    perception 94
anthropocentricity of mathematics 26
*Applicability of Mathematics as a
  Philosophical Problem, The,*
  M. Steiner 26
applicability of mathematics 51, 56–7,
  68
Archimedes, pre-demonstrative methods 82
argumentative stage of justification 82
Aristotle, on explanation 55
arithmetic
  consistency of results 30
  formalization 117–19
  implication of Gödel's theorem 30
  mathematical unassailability 115
  modal structuralism 119–21
  philosophical doubts 115–16

art
  creation 5–6
  invention 4
  Islamic 34
attributive reliability 78
axiomatization, constraints 64–5
axioms 59
  disputes over 110–11
  forcedness 77, 93–4
  Peano arithmetic 65–6

beauty of mathematics 20, 24, 26, 32–3, 37–8
Benacerraf, Paul, on reality of numbers 113
Bohm, David, interpretation of quantum
  physics 28
Bolzano, Bernard, on real definition 93
Borges, *The Library of Babel* 21
Braque, Georges, invention 4
buckyball *141*
  symmetry 141
Burgess, John and Rosen, Gideon 135
  on theories of the world 99

C*-Algebras, axioms 64
calculations, consistency of results 30
calculus, invention 6, 9, 14
Cantor, continuum hypothesis 44
Cappell, Sylvain 139
Cardano, Girolamo, use of imaginary
  numbers 137–8
Cartan, Elie 140
Carter, Howard 4
causal histories 52–3
causal model of explanation 50, 56, 57
certificative stage of justification 82
Changeaux, Jean-Pierre, view of mathematical
  reality 27, 28, 31
chess, comparison to mathematics 136

Cicero, methods of systematic enquiry 81
circle, Euclid's definition 84, 86
Classical Scheme of justification 82–4
Coffa, Alberto, on theories of geometry 104
cognitive command 100, 101–3
  implications of vagueness 107–8
  as prerequisite for objectivity 112
  critique 109–11
  qualifications 107
cognitive shortcoming 101
Cohen, P. J., technique of forcing 6, 9
Columbus, Christopher 4, 6
common sense 114–15
complex number system *see* imaginary
  numbers
concepts
  acquisition 89
  consistency 86–9, 92
  Gödel's view 74–6
  Hermann Weyl's view 90–1
  impossible 85
  Kant's distinction from intuitions 74
  real definition 84–6
    as a practical concern 86–9
    as a theoretical concern 89–91
  Schopenhauer's view 90
  uninstantiated 91
conditioning 80–1
Connes, Alain, view of mathematical
  reality 27, 30
consciousness, physical basis 42–3
consistency of concepts 92
  role of real definitions 86–9
constraints
  in axiomatization 64–5
  in deductive reasoning 65
  role in sense of discovery 63–5
construction
  du Sautoy's symmetrical object 17–19
  of Monster group 6, 8, 14
constructivism 127, 127–8
continuum hypothesis 44
contradiction 85
contrastive questions 53
conventionalism 68
  Gödel's view 75
convergence, as evidence for cognitive
  command 103
Conway's game of Life 10
corpus delicti principle, jurisprudence 84
cosmological role 100
creativity 5–6, 27
  in literature 21

in mathematics 19–20, 21
  musical 21
  unconscious 31
  see also invention
cricket, invention of rules 5
cryptography, application of number
  theory 26
culture, effect on mathematical discovery
  22

'Death and the Maiden' string quartet,
  Schubert 21
Dedekind, J.W.R., on real definition 93
Dedekind–Peano axioms 64
deductive reasoning 65
definitial expansions 123
definitions, value of 84–5
demonstrative methods of investigation 82
dependency relations 57–8
'depth', property of 33, 96
Descartes, René 136
determinate facts 123
Dirac, Paul, pursuit of mathematical
  beauty 33
discovermental methods of investigation 82
discovery 20
  cultural and historical context 22
  definition 4
  distinction from observation 4
  integration across mathematics 23
  mathematical 6, 7–8
  musical 21
  nature of 62
  philosophical perspective 13–15
  Plato's Forms 62
  psychological aspects 11
  role of intuition 31
  sense of 61, 70
    in deductive reasoning 65–6
    Gödel's view 76–7
    illusory nature 67
    implications for mathematical
      reality 68–9
    role of constraints 63–5
discovery/invention question 3, 12, 27, 30–1,
  92
  ancient views 81–4
  implications of consistency 88
  modern views 84–6
disjunctive facts 123
dispositions, effect on perception 79
disputes, and cognitive command 101–8
'Distorted Room' illusion 79

divergent input, in characterization of
cognitive command 105–6, 109

Einstein, Albert, creative imagination 96
electron, symmetry 140–2
eliminative structuralism 121
elliptic curves, solutions to 17
epistemic constraint 100–1
epistemology, relationship to ontology 28
eternal truths 22
Euclid
axioms for geometry 64
definition of a circle 84, 86
Euclidean rescue 104
of logic 105
Euler angles 139
Euler's equation 136–7
surplus value 138–9
Euthyphro contrast 100
evolution of mathematical ability 31–2
'exist', use of term 114–15
existence
of mathematical objects 23, 26
see also mathematical reality
as prerequisite for discovery 14–15
existential axioms, disputes over 110–11
existential facts 123
existential reliability 78
expectation, resistance to, as argument for
physical reality 95
explanation 55–6
causal model 50
inference to the best explanation
(IBE) 53–4
interest-relativity 52–3
mathematical 56–7
mathematical explanations of physical
phenomena 51, 56–7
necessity model 50, 51–2
provision of understanding 49
unification model 50–1
'why-regress' 49–50

facts 122–3
fundamental 124
familiarity, effect on mathematical
perception 80
'felt objectivity' 70–1
see also discovery: sense of
Fermat, Pierre de
Last Theorem 22, 44
proof 24
theorem of primes and squares 24

fermions 142
Feynman, Richard, on electron spin 142
fictionalization 86
Field, Hartry, on logical possibility 67
forcedness
detectability 92–4
felt objectivity as 71
as an indicator of reality 76, 77
mathematical perceptions 78–80
sensory propositions 77–8
forcing, Cohen's technique 6, 9
formal proof
definition of 103
see also proof
formalism, arithmetic 117–18
formalist numbers 130
Forms 62, 74
four-colour-map problem, proof 24
Frege, Gottlob
on conditioning 81
on consistency 87–8, 92
on dependency relations 58
Grundgesetze 71
fundamental facts 124, 125–6

Galileo
gravitational acceleration thought
experiment 52
Il Saggiatore 98
games, comparison to mathematics 136, 144
Gelfrand–Naimark theorem 64
geometry, real definition
practical defence 86–7, 88
theoretical defence 89
'Go', comparison to mathematics 144
Gödel
on concepts 74–5
on feelings of forcedness 71, 93–4
on forcedness of propositions 77–81
phenomenological argument 76–7, 92
Gödel's theorem 30, 66, 92
as argument for mathematical reality 43–4
critique 46–7
and completeness of PA_ω 118
God's eye view 99
good mathematics, quantification 24
gravitational acceleration thought
experiment 52
Gregory, R. L., on Ames' 'Distorted Room'
79
Griess, R., construction of Monster group 8
grounding relation 122–4, 125–6, 131, 132–3
Grounding–Reduction Link 124

group theory
    invention 8
    mathematical reality 95
Grundgesetze, Gottlob Frege 71
Grundlagen der Geometrie, David Hilbert 88

Hadamard, Jacques 140
hallucinations, experience of forcedness 78
Hamilton, W. R., discovery of
    quaternions 63–4, 139
hardness of the logical must, Wittgenstein 63
Hardy, G. H.
    on mathematical reality 61
    A Mathematician's Apology 19–20, 24, 26,
        30–1
Herchel, John, on consistency 87
Heyting, Arend, on intuition 97
Hilbert, David
    Grundlagen der Geometrie 88
    proof-theory 92
historical context, effect on mathematical
    discovery 22
Horgan, Terrence, on independence 97–8
human limitations, role in scientific
    progress 140–1
Hutton, C., on real definition 86
hyperbolic geometry, discovery/invention 10

i
    discovery/invention 6–7, 9–10
    see also imaginary numbers
idealism 99, 127–8
ideas 74
imaginary numbers
    absolute value 137
    discovery/invention 6–7, 9–10
    Euler's equation 136–7
    motivation for introduction 137–8
    surplus value
        quaternions 139
        unitary matrices 139–40
impossible concepts 85
independent reality of mathematics 41, 42–4,
    97–8
    see also mathematical reality
indispensibility 135
inductive reasoning 35–6
inference to the best explanation (IBE) 37,
    53–4, 58–9, 62
Inference to the Best Explanation, Peter
    Lipton 55
inferential error, in characterization of
    cognitive command 105–6

integration of mathematical discoveries
    23, 24
interest-relativity of explanation 52–3
intermediate value theorem 93
intuition 31, 73, 96, 97
intuitions, Kant's distinction from concepts
    74
invention 4–5
    abstract concepts 5
    conclusions 12
    mathematical 6–7, 8, 9–10
    philosophical perspective 13–15
    psychological aspects 11
    see also discovery/invention question
Islamic art 34

jurisprudence, *corpus delicti* principle
    84
justification, Classical Scheme 82–4

Kant, Immanuel 28
    intuitions and concepts 74, 96
    synthetic truths 136
Kant–Quine thesis, objectivity 99, 107
Kepler, Johannes, laws of planetary
    motion 140
Kitcher, Philip 57
Klein, Felix 139
knowing, distinction from understanding 49,
    55
knowledge, philosophical theory 116
Kreisel, Georg 66, 135
Kronecker, Leopold 23

Laguerre, E. N., on matrices 141
Lakatos, Imre, 'monster-barring' 107
language of the universe (Galileo) 98–9
latent information 136, 138, 143
'laws', mathematical 57
Leibniz, G. W. F.
    on definitions 84–5
    on impossible concepts 85–6
    invention of the calculus 6
Leslie, John 89
Library of Babel, The, Borges 21
Lie groups 144, 145
Life, game of 10
Lipton, Peter 55
    on inference to the best explanation 37
literary creativity 21, 23
logic, objectivity 105–7
logical consequence 65–6
    'felt objectivity' 70–1

logical possibility 66–7
logic-choice 106
Lorentz group 143

Mandelbrot set 6, 30
materialism 27, 28
  unsatisfactoriness 29
mathematical objects 61–2
  Forms 62, 74
  minimal realism 114–16
  qualified realism 132–3
    modal structuralism example 119–21
    reductionist examples 117–19
  reality 113–14
    assessment of reductionism 128–31
    reductionist proposal 124–6
  reducibility 121–2
  scientific confirmation 63
mathematical perception 80–1
  distinction from sensory perception
    78–80
mathematical reality 20, 27, 61, 92
  analogies with physical world 29–30
  and evolution of mathematical ability
    31–2
  'forcedness' as an indicator 77–81
  Gödel's theorem as argument for 43–4
    critique 46–7
  Gödel's view 76–7
  group theory 95
  implications of mathematical
    discovery 68–9
  intuitive perception of 31
  metaphysics 28–9
  theories as evidence 62
  unreasonable effectiveness of mathematical
    beauty 32–4
mathematical thinking, role of intuition 31
mathematical understanding 50
Mathematician's Apology, A, G. H.
    Hardy 19–20, 24, 26, 30–1
mental life, materialist view 29
metaphysical disputation 29
metaphysics 28–9
  objectivity 111
mind–brain relationship 28–9
minimal realism 114–16
modal structuralism 119–21
modal structuralist numbers 130
monetary reductionism 125
Monster group, discovery/invention 6, 8, 14
'monster-barring' 107
motivation 20, 23–4, 144–5

for introduction of imaginary
    numbers 137–8
music, comparison to mathematics 23–4
musical appreciation 33
musical creativity 21, 23
musical discoveries 21

nature/nurture debate 24–5
necessity model of explanation 50, 51–2, 56
neutron stars, PSR B1913+16 system 44
Newton, Isaac
  invention of the calculus 6
  use of analogy 36
noēsis 73
noetic realm hypothesis 73
  argument by analogy 29–30
  evolution as argument for 31–2
  unreasonable effectiveness of mathematical
    beauty 32–4
non-Euclidean geometry,
    discovery/invention 7, 10, 20–1, 23
non-explanatory proofs 50–1
  gravitational acceleration thought
    experiment 52
noumena 28
nucleons, SU(2) symmetry 142
number systems 129–30
number theory, application to
    cryptography 26
numbers
  Platonist view 124
  reality 113–14, 115–16, 132–3
  formalist view 119

objectivity 97–8, 135–6
  cognitive command as prerequisite 112
    critique 109–11
  compromises 107
  deductive reasoning 65–7
  'human' influences on theorizing 99
  Kant–Quine orientation 99
  of logic 105–7
  as a metaphysical concept 111
  Wright's account 100–1
    cognitive command 101–8
observation, distinction from discovery 4
Ocken, Stanley 139
omega sequences 120–1
ontology, relationship to epistemology 28
optical illusions
  Ames' 'Distorted Room' 79
  experience of forcedness 78
ordered pairs, complex numbers as 9

patterns of events, mathematical
explanation 51, 56
Peano arithmetic 117–19
Peano axioms 65–6
Penrose, Roger 29
on imaginary numbers 138
on objectivity 135–6
perception, consistency 29–30
perception-like experience of mathematical
concepts 76
Permanence, Principle of 80
persistence 126
personal experience, materialist view 29
phenomena 28
phenomenology of mathematics 61
Gödel's argument 76–7
physical behaviour, dependence on
mathematics 41–2, 44–5
physical explanation
causal model 50
necessity model 50
physical phenomena, mathematical
explanations 51, 56–7, 68
physical realism 95–6
physical world, analogies with
mathematics 30
physicalism 115
physics
relationship to metaphysics 28
search for beautiful equations 32–3
Π_1-sentences 44
Picasso, Pablo
creation 6
invention 4
planetary motion, Kepler's laws 140
Plato, Forms 62, 74
Platonism 3, 23, 124, 135
degrees of 43–4
independent reality of mathematics 41,
42–4
physical behaviour, dependence on
mathematics 41–2, 44–5
Playfair, J. 86–7
Poincaré, Henri, intuitive perception
31, 96
pre-demonstrative investigation 82
primes
discoveries 6
Fermat's theorem 24
Riemann Hypothesis 22, 23
Principle of Permanence 80
problematic investigations 83
Proclus, 'ordering' of Propositions 82–4

proof
cognitive command 101–8
creativity 23
discovery/invention 7, 10–11, 65
formal, definition of 103
motivation 23–4
non-explanatory 50–1
gravitational acceleration thought
experiment 52
requirement of axioms 110
Protagoras 99
PSR B1913+16 neutron-star system 44
Putnam, H. 135
pyramids, Egyptians' calculation of
volume 145

quadratic formula, discovery 7
qualified realism 114, 116–17, 125, 127,
132–3
modal structuralism example 119–21
reductionist examples 117–19
quantum physics
counterintuitive nature 30
interpretation 28
reality 95
quarks, SU(3) symmetry 142
quaternions 139
W. R. Hamilton's discovery 63–4
Quine, W. V. 135
quintic, insolubility 7

Ramanujan, Srinivasa, intuitive perception
31
real definition 84–6, 92
Bolzano's views 93
as a practical concern 86–9
as a theoretical concern 89–91
realists 28
reality
dimensions of 27
see also mathematical reality
recursively enumerable theories 118
reductio proofs 50
gravitational acceleration thought
experiment 52
reductionism 117, 121–2, 124
assessment 128–31
reflective equilibrium 106–7, 112
relevance 128
reliability, existential and attributive 78
research, psychological aspects 11
Resnik, Michael 104
on wide reflective equilibrium 106

richness, analogy between mathematics and
  physical world 30
Riemann Hypothesis 22, 23
Rosen, Gideon, on dependency relations 57–8
rotations
  representation by complex numbers
    quaternions 139
    unitary matrices 139–40
  symmetry of electron 140–2
Russell, Bertrand, on axioms of set theory 94

Schopenhauer, on nature of concepts 90
Schubert, 'The Death and the Maiden' string
  quartet 21
science
  limited aims 28
  richness of physical universe 30
scientific explanation 55–6
scientific progress, role of human
  limitations 140–1
scientists, realism 28
semantic consequence 66
semantic indeterminacy 130
sensory perception
  analogy to forcedness 94
  distinction from mathematical
    perception 78–80
sensory propositions, forcedness 77–8
set theory, axioms 77, 93–4
SL(2,C) matrices 143
social constructivism 128
'sporadic' finite simple groups 30
Steiner, Mark 57
strong reductionism 124
structuralism 119–21
SU(2) symmetry 139–40
  buckyball 141
  electron 140, 142
  nucleons 142
SU(3) symmetry 142
supervenient facts 123
surplus value 138, 144
  imaginary numbers
    quaternions 139
  properties of buckyball 141
  unitary matrices
    139–40
    SL(2,C) 143
    SU(2) 141–2
    SU(3) 142

surprise, analogy between mathematics and
  physical world 30
symmetry
  designs of Alhambra Palace 18, *19*, 34
  du Sautoy's construction 17–19
synthetic truths 136

Taylor series expansion 68
theorematic investigations 83
theories, discovery/invention 6, 14–15
theory development
  axiomatizations 64–5
  constraints 63–5
  discovery of quaternions 63–4
  as evidence for mathematical reality 62
Thompson, J. J. 4
thought, mind–brain relationship 28–9
three 'worlds' 41, *42*
training, effect on mathematical perception
    80
transfinite numbers, discovery/invention 14
Truth and Objectivity, Crispin Wright 100–1
  cognitive command 101–8

understanding
  distinction from knowing 49, 55
  mathematical 50
unification model of explanation 50–1, 56, 57
uninstantiated concepts 91
unitary matrices 139–40
unreasonable effectiveness of mathematical
    beauty 33, 37–8
utility of mathematics 20, 23, 26

vagueness, implications for cognitive
    command 107–8

Waismann, Friedrich, on Taylor series
    expansion 68
warfare, use of mathematics 26
wave/particle duality of light 30
Weyl, Hermann, on nature of concepts 90–1
why-questions, contrastive form 53
'why-regress' 49–50
wide reflective equilibrium 106–7
Wigner, Eugene, on mathematical beauty 33
Wittgenstein 63
  on mathematical proof 67–8
Wright, Crispin, *Truth and Objectivity* 100–1
  cognitive command 101–8